情境分類 × **換句話說** × **練習試題**

讓本書用最多樣的主題搭配最精準的用字，

幫助你寫出最讓人滿意的英文 E-mail！

Preface 前言

大量的數位資訊，除了拉近了世界各地的距離，更大量衝擊著目前的商業行為。許多貿易公司僅僅透過電腦及電話，就能創造數十億的營業額，這就證明了語言╳數位，就能營造出驚人的經濟奇蹟。

在我過去幾年的商業經歷中，我發現語言已經不是一種知識，而是一種能力，更是一個讓自己被看見的最好工具。但是許多身邊的朋友，包括同事，當面臨語言能力不足時，常常尋求坊間英語補習班的協助，每天甚至花上兩個小時努力練英文，這麼多的努力卻無法立即地反應在工作成效上，他們的挫敗常常令我感到同樣的沮喪。

憑藉著我從小對於英語學習的高度興趣，除了結合我的英語教學經驗，最後更是結合了我在商場上的實際經歷，我於今年年初開始著手編寫我的第一本英語學習工具書《深度解密！一次就上手的超實用職場英文 E-mail 即戰手冊》，並且很開心能夠與捷徑文化一起合作出版這本作品。

在這本書中，我不只與各位分享如何快速又得體地寫出一封簡潔的英文 E-mail，更利用這些 E-mail 實例，同步分享英文文法的學習關鍵及文法概念，希望讓所有對英語有興趣、有學習需求的讀者，可以用最少的時間學會最大的能力。

我誠摯地希望這本書的內容不會讓各位失望，也希望我的用心可以給各位最大的幫助；如果你想讓自己不再成為一個可以被取代的員工，相信《深度解密！一次就上手的超實用職場英文 E-mail 即戰手冊》應該可以給你最大的幫助。

本書使用方式

密技1 以必備暖身技打底，告訴你寫英文郵件的撇步

在進入寫信和回信的練習之前，本書先在 Part 1 告訴讀者英文書信的格式、稱謂、敬語，以及其他繁複的「眉眉角角」。編排一目瞭然，輕鬆掌握職場 E-mail 需注意的語氣問題，才不會讓對方一看到主旨就不想點開來看！

密技2 情境式分類的單元，不怕找不到你要寫的主題

本書收錄多元豐富的實用範例，除了包含廣泛的問候合作等主題，還有收入各個部門的細項內容。一本書就滿足你所有職場上實際會面臨的情況，讓你不用再為了寄信、回信而苦惱。

密技3 字詞句型換句話說，教你依照實際情況變化寫法

本書的每單元不但有參考的寫信範例，還標示列舉可以變化寫法之處，詳細講解句子所使用的文法概念、句型結構、片語意義，以及替換單字的不同用處，教你在嚴守標準的書信格式下，也能靈活替換表達方式，延展書寫彈性。

密技4 隨附「馬上練習」，打鐵趁熱學習效果立見真章

此部分為填入關鍵句與獨立完成段落的引導式寫作方式，讓讀者在應用過程中確實了解運用的方法。本書也附上精選參考解答，並列出模板句型，讓讀者可以擁有比較的範本，還能溫故知新，活用各種不能不會的職場英文 E-mail 寫法！

Contents 目錄

Part 1 英文E-mail 必備暖身技

沒有正確格式，專業度就無法令人信服！

Unit 01 英文 E-mail 不能不懂的重點格式說明

隨著時代的變遷，E-mail 已大量取代傳統的書面英文書信；加上網路科技所帶來的即時訊息傳遞與同步進行的視訊會議等高校傳輸，點開 E-mail，專業度便分曉！專業格式化，無形力量大。本單元將一步一步帶領你寫出專業E-mail，讓你的專業度即時被看見！

完全拆解E-mail 格式

STEP1 信件訊息

一封商業E-mail可能包含的資訊有：收／寄件人的姓名（name）、郵件地址（email address）、信件主旨（email subject）、公司市話（telephone number）、公司行號（company name）、職稱抬頭（position title）、起頭／結尾敬語（complimentary close）等等。

STEP2 內文格式之要素解讀

▶ **要素① 起頭的敬語**

稱謂一定要先確認其性別與姓名，如果無法得知收件人姓名則使用公司職位抬頭來稱呼。

▶ **要素② 正文篇幅與內容**

內文則如同中文的起承轉合為基底。商業書信是為了以最少的時間發揮最大效益，因此建議一開頭的問候語之後即可簡單表明今天的來意與總結陳述。之後，再分段落做詳細說明，段落一定要分明，請勿只打一段，最好每段呈現一個重點，才不會通通擠在一起而容易漏掉某個點。倘若只是很簡單的進出口貿易書信，則沒有必要做多段落的分劃，請務必注意！

再來，因為 E-mail 不像口語或表情可以幫我們直接傳遞感覺給對方，因此如欲告知對方重點部分則可全部以大寫字顯示，但盡量用於關鍵字上面，避免泛濫使用。其他呈現重點的方式也可以用粗體字或是highlight（重點螢光上色）。

▶ 要素③ 結尾敬語

如同中華文化的傳統，即便是結束離去都要有祝福對方的話語，並且依據與對方的狀況和了解而給予當下最適當的祝福。一句話可以溫暖人心又不花錢，何樂而不為呢？現在就開始養成好習慣，從小地方開始做好國民外交吧！

情況	正式用法（商業書信）	非正式用法（熟識友人與關係）
常用於對上司或較敬重的人	Best Regards, Best Wishes, Sincerely yours, Truly yours,	All the best! Take care! Good luck! Love,
一般情況	Regards, Thanks,	Warmly, Cheers!（英式用法）
通用情況	Have a nice day! / Have a good day! / Have a great weekend!	

▶ 要素④ 署名

當信件內容與結尾敬語都完成了，接下來就是自己的全名（通常是第一次接洽或新開發的時候）、自己的職銜與公司相關資訊等的提供，以便他人輕鬆搜尋與資料整理。切記資料一定要完整，如果是一封開發信更不可漏掉這麼重要的訊息！否則客戶想做直接聯繫都沒辦法，在無形中白白喪失一個合作機會。

另外，還需注意信件的格式，完整呈現如下：

To（收件者）
Cc（Carbon Copy 信件副本）
BCC（Blind Carbon Copy 密件副本）

Subject（信件主旨）
Attachment（附件）

Body（內文）

Complimentary Close（結尾敬語）

Name（送信者名稱）
Position（公司職稱）
此處亦可嵌入公司的小型識別ID / Logo

Company Name（公司全名）
Company Address（公司地址）
Company Tel. No.（公司電話號碼）
Company Fax No.（公司傳真號碼）
Company Website Address（公司網址）

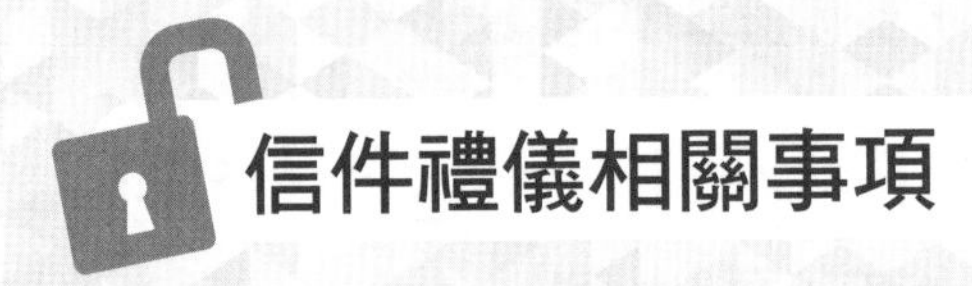

信件禮儀相關事項

▶ 全盤檢視

信件精簡有力不要有過長的陳述句子，商場上時間就是金錢；內容要一次說清楚、一次到位，最好不宜超過三段。文字不須要使用艱深的單字，E-mail 是重溝通不是在考托福、玩文字遊戲。在按下SEND鍵之前，請先再閱讀一次自己的信件，並查看是否有拼錯字或句義不完整的字句。

▷ 附件檔案

Attached (with)... 就是最簡短有力的說明

請在附加檔案前檢查是否附上正確且相關的檔案，並做病毒的掃瞄偵測以防對方電腦系統中毒。如遇對方來信顯示中毒跡象，為使彼此損害減到最低，應立即反應給對方。

▶ 回信

商業往來信件應每天定時閱讀並做回應。即便無法在第一時間做回覆，至少可以告知已收到信件，以避免對方等候多時然後繼續寄信件而將信箱淹沒。

▷ 出差與休假

長時間無法讀取信件時，記得事先做好系統自動回覆的設定，可以的話可以告知緊急狀況的代理聯絡人讓彼此都安心。

▶ 轉寄信件

不論於公於私，信箱常會收到與公事不相干的訊息，盡量避免抱持著分享的心情而轉寄別人不必要也不需要的信件。

無趣的信件標題，刪信率高達85%!

Unit 02 英文E-mail 不能不學的重點下標法則

隨著產業別的不同，打開公司E-mail 信箱就有一堆看不完的信件，有提案信、詢問信、開發信、報價信、客訴信等等，再外加上莫名的廣告信，你該如何在茫茫信海中讓人辨識出你信件的輕重緩急？請務必注意，以下所提出的都是為了幫助你達到最快也最有效率的下標題法則！

下標重點

下標題前的基本概要

首先，一封「可以被順利寄出」的 E-mail 除了有明確表達的內文之外，最主要的基本元素就是下列兩項：

To（收件人） 你想寄信的對象	Subject（信件主旨） 你想要傳遞的訊息主旨
在輸入之前請務必確認郵件地址無誤，否則就算標題再好、內容再棒，也是「信」沉大海！	「己所不欲，勿施於人」抱持這個信念來下標題就對了！一針見血的精簡度點出來信所為為何！
EX: anna@yahoo.com.tw	EX: Request for S/S 09 catalogues

何謂一針見血的標題法則

商業書信因領域與生意模式的不同，出現了無法數計的信件類型。相信你也曾經歷過在追蹤前一位同事的工作進度時，需要去查詢E-mail 往來紀錄的狀況，是否都能即時找到「關鍵信函」呢？Subject（主旨）不只要精簡更要具體（務必盡量指名其產品型號或其規格與重要日期），因為那是為方便日後彼此做「追蹤」或作為「依據查詢」時的重要指標！平時的小小一個動作都會影響著日後做事的效率！

活靈活現！（常見範例列舉）

情況	正式用法（商業書信）
關於庫存問題	Question(s) about Product inventory
關於2009年財報問題	Questions about 2009 sales report
詢問2009年春夏系列產品	Inquiry about 2009 S/S collection
詢問宏基小筆電Aspire One Sample	Inquiry for Sample Acer Aspire One
詢問勞力士錶款11234 型號	Inquiry about Rolex watch #11234
產品11x03 的備材須求	Request for Spare parts of #11x03
437型號產品備材須求	Request for Spare parts of model #437
10/25提貨單編號55204	10/25 B/L #55204
手機23型號出貨通知	Cell phone#23 Shipment Notice
11/03樣品出貨	11/03 Sample Shipment
至4/18止的欠貨明細單	Pending list as of April,18th
緊急- 修正後的訂單 （讓收件者知道迫切性的明顯標示）	Revised Order list <URGENT!!!> 或 ***URGENT***
ASUS 2010年價格表單	ASUS 2010 Price List
客訴信件統整概要與建議	Summary of Customer complaints &Suggestions
2010年春季行銷企劃	2010 Spring Marketing Planning
12/10 業務部會議議程	12/10 Meeting Agenda of Sales Department
回覆：11/03 樣品出貨	RE：11/03 Sample Shipment
轉寄：12/10 會議議程	FW：12/10 Meeting Agenda

◆附註：在重要的標題字眼上以大寫為起始可幫助視覺上重點吸睛的效果。

現學現賣

(A) 根據下表內文，找出最適合的主旨

(a) Customer complaints - Product issue #11x05-00256
(b) Product Issue and Replacement on #32165
(c) Special Offer of A/W 09
(d) Urgent Shipment Invoice #2314
(e) Inquiry about Product Issue of #11x05-00256
(f) Shipment Notice of Model #2314
(g) Repair Issue #32165

(1) ______We are here to inform you that several of our customers have come back to us complaining about product #11x05-00256. Its watch chain cannot be well-adjusted. Please help us to find out the quick solution. Thank you!

(2) ______ Due to the recent financial crisis, we all suffered a lot; therefore, we would like to make a special offer on A/W collection 09. Please see the attachment where you could find out the new price list.

(3) ______ We are pleased to inform you that your order of model #2314 will soon be released within 2 days. Please confirm the final check of the invoice, as attached. Thank you!

(4) ______ Product code 32165. Please see the photo in the attachment. There are several irremovable stains and bad electroplated spots, and we would like to send those products back and have other replacements. Thank you!

◆**解答：**(1) a，相較於e 的標題只是告知有產品問題的須求，a 是以客訴為主標，讓重視客戶的企業迅速處理危機／(2) c，簡明直接切入企業在乎的重點／(3) d，相較於f 的標題，因為馬上要出貨的程序繁瑣，所以打出緊急的字樣可以讓對方盡快確認與處理才不會影響預定的出貨時間／(4) b，相較於g 則是更完整的說明。

(B) 依據下列短文，給予最貼切、吸睛的標題！

(1) ______________________________

We are interested in your products and probably would like to cooperate with you. Please let us know if we could get several pairs of your model samples. We would appreciate it if you could send them within a week. Thank you!

(2) ______________________________

We've noticed that there might be new inquiries in your local market, and we have the best offer to meet your current needs. Attached is the up-to-date price list and catalogues of 2010. Please let us know if you need any further information. Thank you!

(3) ______________________________

We are recommended by others from Sunny Corp. that you have the best quality of sunglasses in Taiwan. The thing is that we will be having a huge inquiry for our annual order. Therefore, please contact us with your best quotation. Thank you!

(4) ______________________________

We have received large inquiries on our S/S 2010 collection among our agents around the world within this week. Therefore, in order to meet your seasonal order quota and availabilities of each collection, please make the order as soon as possible. Thank you!

◆解答：(1) Request of model samples / Prospective Business-Model sample wanted (2) News! Price list and catalogues of 2010 / Best Choice and Price of 2010 (3) Quotation Request of annual order / Best quotation wanted (4) Notice of Stock Order / Alert! Short in-stock

Part 2

英文E-mail實例大全——主題篇

Chapter

- 01 預定約會篇（Appointment）
- 02 合作業務篇（Establish Cooperation）
- 03 專案進行篇（Project Proceeding）
- 04 商業交易篇（Trade/ Trading）
- 05 提出要求／抱怨篇（Asking and Complaint）

Unit 01 詢問受訪意願

寄信時可以這麼說

Dear Mr. Anderson,

I am the chief editor from Newest Magazine and I am writing to **make a request for** an interview with Mr. Beckham, the CEO of your company. Mr. Beckham has always played an important role ① among the nation ②, so it would be a great honor and a **beneficial** ③ experience for us to get this chance. The interview will be **focused on** Mr. Beckham's **perspective** on the economical situation of the world, and it would lost about ④ an hour. Please let us know if Mr. Beckham **grants** us the opportunity to interview him. Any time which is convenient for Mr. Beckham is acceptable for us.

I am hoping for a **favorable** reply. Thank you very much.

Sincerely,

Laura Fang

親愛的安德森先生：

我是嶄新雜誌的總編輯，在此**提出**採訪貴公司總裁——貝克漢先生**的請求**。一直以來，貝克漢先生都扮演著國內②重要的角色①，所以若能得到這次機會，對我們來說一定是個非常榮幸且**受益良多的**③經驗。採訪的主題將會**聚焦於**貝克漢先生對未來世界經濟局勢的**看法**，為時大約④一小時。請告知我們貝克漢先生是否**同意**受訪。任何貝克漢先生方便的時間，我們都可以接受。

希望能有個**肯定的**答覆。非常感謝您。

方蘿菈 謹上

換句話說

① **be of great influence 有著重要的影響力**
→形容重要的人物除了如郵件中的「重要角色」（an important role）以外，還可以形容此人「有著重要的影響力」（be of great influence）。

② **nationwide／internationally 國際上**
→「among + 範圍」可表示「在某個範圍內」，例如：among the co-workers、among the family members等。若想用簡潔一點的方式表達，可以直接用nationwide一詞。另外，若是表達「國際上」可用internationally。

③ **informative 增長見聞的／valuable 可貴的／profitable 有益的**
→通常讚賞一個人的演講、報告等表現佳，可用「受益良多的」（beneficial）、「增廣見聞的」（informative）、「可貴的」（valuable），或是「有益的」（profitable）等表示。

④ **be approximately／be around**
→「大約」的表示方法除了常用的about，還可以用approximately和around表示，後面直接加上時間即可。

- ♥ **beneficial** *adj.* 受益良多的
- ♥ **perspective** *n.* 看法
- ♥ **grant** *v.* 同意；授予
- ♥ **favorable** *adj.* 肯定的；贊同的
- ♥ **make a request for...** 提出……的要求
- ♥ **focused on...** 聚焦於……

回覆時可以這麼說

Dear Ms. Fang,

I am happy to inform you that Mr. Beckham is delighted to accept your interview. Mr. Beckham will **take a business trip** to New York ① from June 17 to July 3 ②, so I will have to ③ **arrange** the interview on this Friday, June 16, at 9:00 a.m. in Mr. Beckham's office. Should there be any problem, please let me know immediately.

By the way, Mr. Beckham also wanted to express his excitement of having this chance to share his own take on global economy in your interview. Since Newest Magazine is a **reputable** company, I believe this would be a **win-win** situation for both of us ④.

Thank you for inviting us.

Sincerely,

James

親愛的方小姐：

很高興通知您，貝克漢先生願意接受您們的採訪。貝克漢先生在六月十七日到七月三日②將到紐約①**出差**，因此我必須③將採訪**安排**在本週五，也就是六月十六日的上午九點，地點是在貝克漢先生的辦公室。如果有問題，請您立刻知會我。

順帶一提，貝克漢先生也想向您傳達他對於這次能夠在您們訪談中分享自己對國際經濟想法之機會的興奮之情。因為嶄新雜誌也是**享有聲譽的**公司，相信這對我們雙方來說會是**雙贏的**局面④。

謝謝您的邀請。

詹姆士 謹上

換句話說

① **our London Branch 我們的倫敦分公司**

→這裡除了可以放城市、地區或國家名以外，如果公司在別地有分公司，也可以用於此處，直接以「地名 + Branch」表示。

② **between June 17 and July 3**

→「時間範圍」的表達方式除了用from...to「從……到……」以外，還可以用between...and「在……和……之間」。要注意的是若是坐落在特定的日期上，介系詞需要用on。。

③ **I have no choice but to 我不得不……**

→have no choice but to用來表示情況很艱難，所以被迫做出的選擇。在口氣上來説相對強烈。

④ **both of us would benefit from this interview**
我們雙方都會在此次的採訪中獲益

→在商業配合中，常常需要衡量收益、虧損，以做出簽約、合作等結果。在郵件中，方小姐提到「雙贏的局面」，表示採訪對雙方的知名度及收益方面都會有所提升，所以可以用win-win表示。除此之外，也可以用both benefit from...表示。

♥ **arrange** *v.* 安排
♥ **reputable** *adj.* 名聲好的
♥ **win-win** *adj.* 雙贏的
♥ **take a business trip** 出差

馬上來練習吧！

想要詢問受訪意願時該怎麼說呢？

1.	開頭稱謂	Dear Kevin,
2.	問候句	I am writing to ______________________________ ____ ____.
3.	信件主要內容	To provide you with ___________________________, ________________. Could we ________________? If ______________, _____________________________ ____ ____ ____.
4.	結尾問候句	Expecting to meet you soon.
5.	署名	Yours sincerely, Alice

參考解答 *Answer*

Dear Kevin,

I am writing to follow up on our telephone conversation regarding our new project. To provide you with more detailed information, I would like to pay you a visit. Could we make an appointment sometime between June 17 and 19? If it is not convenient with you, would you be kind enough to let me know when you will be available?

Expecting to meet you soon.

Yours sincerely,

Alice

中譯

親愛的凱文：

我是寫信來繼續我們在電話上談論關於本公司新企劃的事情。為了提供您更詳細的資料，我想要拜訪您一趟。我們能不能在六月十七日和十九日這段期間約個時間呢？如果這時間您不方便，能不能請您告訴我您何時有空？

期待能盡快與您見面。

愛莉絲 謹上

句型這樣替換也可以

1. **I am the assitant editor from...** 我是……的助理編輯。
2. **I am writing to inquire...** 在此詢問……。
3. **The interview will revolve around...** 採訪的主題將會集中在……

Unit 02 詢問會面時間

寄信時可以這麼說

Dear Samuel,

Since you mentioned that you would be **unavailable** for meeting me on this Friday, I would like to know if it is possible ① for me to **reschedule** you for an **appointment** on either May 10 or 11, at 11:00 a.m. Please let me know if either is convenient ② with you. I know that both of us are really busy recently, but this project matters a lot ③ for me. The manager has returned my proposals for many times, but this time it should be the final version, so this is really an **emergency** ④.

I'll be looking forward to your reply **at your earliest convenience**.

Thank you very much.

Best regards,

Danielle White

親愛的山繆：

由於您提到這週五**無法**與我碰面，我想知道是不是可以①跟您**重新安排約定時間**，也就是五月十日或十一日的上午十一時。希望這個時間您方便②見面。我明白我們最近都很忙，但是這個專案對我來說很重要③。經理已經將我的提案多次退回了，而這次一定要定案，所以這次真的很**緊急**④。

希望**您一有空**就能回覆我。

非常感謝您。

祝福您

丹妮兒．懷特

換句話說

① **there would be a chance 我有機會能……**

→郵件中的possible表示「可能的」，類似的表達方式還有名詞chance和opportunity（機會）。不過，雖然兩個替代詞都譯為「機會」，但後者通常會用來表示「某種（正面的；好的）機會」。

② **acceptable 可接受的**

→通常和他人詢問會面時間，會用的詞除了convenient以外，常用的還有available和acceptable，可於此替換使用。

③ **is of great importance**

→「be of + N」片語可表示某物的「某項特徵」。依照郵件想表達的「專案很重要」意思，便可以用is of great importance表示。

④ **an urgent situation**

→和emergency同義的常用單字有urgency一詞，都是表達「緊急事件」。除了可以直接將原句中的emergency替換為urgency外，還可以轉化詞性，用形容詞片語an urgent situation表示。

♥ **unavailable** *adj.* 沒空的；無法利用的
♥ **reschedule** *v.* 重新安排時間
♥ **appointment** *n.* 約定
♥ **emergency** *n.* 緊急事件
♥ **at your earliest convenience** 你一方便時

回覆時可以這麼說

Dear Danielle,

I would like to thank you for your understanding.

I'm afraid that I will have to attend a **conference** in our **headquarters** on May 10, so May 11 will be better for our meeting ① . I will be expecting you in my office ② next Wednesday at 11:00 a.m. to discuss your **proposal**.

As to your **concern**, I'll have to let you know that there's nothing to **be worried about** this time. Your last **version** of the proposal is almost perfect ③ ; there are only a few details ④ to be fixed.

Regards,

Samuel

親愛的丹妮兒：

謝謝你的體諒。

五月十日我恐怕必須到總公司出席一場**會議**，所以我們五月十一日這天在**總部**碰面會比較好①。下週三早上十一時，我會等你到我的辦公室②來商討你的**提案**。

至於你的**焦慮**，我必須勸你這次不必**擔心**。你上一份**提案**已經近乎完美了③，只有一些細節④需要修改而已。

山繆 謹上

換句話說

① **we'd better meet on May 11**
→had better一詞通常用於祈使句，表示「要求／請求人做某事」，口氣較無郵件中的句子委婉、禮貌。用法為「had better + 原形V」，通常會縮寫為「-'d better」。

② **I will meet you in my office...**
→郵件中的原句expecting一詞之意為「期盼；預料」，所以此句表示「未來（下星期三早上11點）將會在辦公室見面」。而同樣表示未來的動作的用法還有「will + 原形V」，因此這句可改為I will meet you in my office。不過，口氣上而言，原句的用法會較為客氣。

③ **You've nearly done great on your last version of the proposal.**
→此句想表達的是「提案已經近乎完美了」，除了郵件中的「N（最新版提案）be almost perfect」之外，還可以調換句型，使用替代句「aready done a great job on + N（最新版提案）」。

④ **statistical errors 數據錯誤／spelling mistakes 拼字錯誤**
→商業溝通中，提案常出現的用詞detail（細節）涵蓋範圍很廣。其可能表示數據statistics，或者spelling，這裡的替代用法的兩種「錯誤」error和mistake是常用搭配法，因此通常不會交錯使用。

單字片語急救包

- ♥ **conference** *n.* 會議
- ♥ **headquarters** *n.* 總部；總公司
- ♥ **proposal** *n.* 提案；提議；求婚
- ♥ **concern** *v./n.*（使）擔心；（使）憂慮
- ♥ **version** *n.* 版本
- ♥ **be worried about** 為……擔心

馬上來練習吧！

想要詢問會面時間時該怎麼說呢？

1. 開頭稱謂

Dear Linda,

2. 問候句

Thanks ______________________________
____
______________________________.

3. 信件主要內容

I am pleased ______________________________.
______________________________,
____
______________________________.

4. 結尾問候句

Since I will ______________________________
_____ ______________________________
____
______________________________.

5. 署名

With regards,

Samuel

參考解答 *Answer*

Dear Linda,

Thanks for your letter of June 12. I am pleased to meet you this Friday, June 17 at 10 a.m. Since I will be attending an important conference that afternoon, I hope you don't mind meeting me in the morning.

I will be expecting you in my office.

With regards,

Samuel

中譯

親愛的琳達：

謝謝您六月十二日的來信。我很樂意在這個週五，也就是六月十七日的上午十點與您見面。因為我那天下午要出席一場重要會議，所以希望您不介意上午來跟我碰面。

我會在辦公室靜候光臨。

山繆 謹上

句型這樣替換也可以

1. **Please let us know if...** 請告知我們是否……
2. **I am happy to inform you that...** 很高興通知您……
3. **I would like to know if it is possible for me to...** 我想知道是不是有可能……

Unit 03 會議延遲

寄信時可以這麼說

Dear Mr. Spears,

I regret to inform you that the **franchise** organization **seminar** ① we scheduled on March 22 will be **postponed** ② till March 29, at 7:00 p.m. due to some unexpected **incidents** ③. The seminar, as planned, ④ will be held ⑤ in the Third Conference Hall in Fu Xing Building. We are sorry to have caused such inconvenience. Please accept our sincere apology. If there's anything we could do to improve upon the whole session planning, feel free to let us know.

Thank you very much for your understanding.

Yours faithfully,

Stephanie Chou

親愛的史皮爾斯先生：

很抱歉通知您原定在三月二十二日舉辦的**加盟說明會**①，因某些突發**事件**③將**延遲**②至三月二十九日下午七時舉行。說明會的地點將按照原定計畫④，在復興大樓的第三會議廳舉行⑤。我們很抱歉造成如此不便。請接受我們誠摯的道歉。若在整體會議策劃上有能改進的地方，請不吝指教。

非常感謝您的體諒。

史蒂芬妮．周 謹上

換句話說

① **press conference 記者招待會／**
the Energy Recycling Seminar 能源再生研討會／
the Shareholders Meeting 股東會

→郵件中將舉行的是加盟說明會 franchise organization seminar，其他常出現的會議還包含記者招待會press conference、能源再生研討會the Energy Recycling Seminar，和股東會the Shareholders Meeting。

② **adjourned／delayed／suspend**

→延遲的英文有許多，其中除了郵件中的postpone以外，常見的還有adjourn、delay和suspend等，只要在後面加上「until / till + 時間」即可。另一種用法表示「延遲持續的時間」則用「for + 時間範圍」。

③ **the typhoon 颱風／some internal factors 某些內部因素**

→造成事件延遲的原因可能有很多種，郵件中的原因為某些突發事件some unexpected incidents，其他還有可能是因颱風the typhoon，或是某些內部因素some internal factors等。

④ **As the original plan, the seminar...**

→郵件中的原句要表達的是「說明會照計畫在原地方舉行」，除了用plan的完成式as planned表示「和已計畫的一樣」，也能用plan的形容詞片語the original plan替換。

⑤ **take place**

→「舉行；舉辦」的用法除了郵件中的be hold...以外，還可以用take place代替。

單字片語急救包

♥ **franchise** *n.* 特許經銷權
♥ **seminar** *n.* 研討會；專題討論會
♥ **postpone** *v.* 延遲
♥ **incident** *n.* 事件

回覆時可以這麼說

Dear Stephanie,

I got your e-mail in which you notify us of the postponement of the seminar. Since I will **be on a business trip** to Los Angeles ① from March 25 to March 30, it would be impossible for me to ② **attend** the seminar on March 29 personally ③. So ④ I will have Mr. Robinson, our Business Manager, **represent** me. He is a **reliable** employer, so there is nothing to be worried about.

Regards,

Simon Spears

親愛的史蒂芬妮：

我已經收到您通知研討會將有所延遲的信件。因為我在三月二十五日至三月三十日這段時間要到洛杉磯**出公差**，將無法②親自③**出席**三月二十九日的說明會。所以④我將請我們的業務經理——羅賓森先生**代表**我出席。他是一個**可靠的**職員，所以沒有什麼需要擔心的事。⑤

賽門．史皮爾斯 謹上

換句話說

① **out of the town 到外地去／on holiday 休假**

→短暫離開一地常見的原因有：出差be on a business trip、出城be out of town、休假be on holiday、到……出差be on business trip to...等。用法都是在前加上be介系詞。

② **I am unable to...**

→英文中的書信用語，常常為了要正式一點，而將主動轉被動，或是使用「需主詞it」表示，郵件中這句it would be impossible for me to...就是屬這種用法。其實這句也可以簡單地用I am unable to...句型表示。

③ **in person**

→personal一詞表示「個人的；私人的」，因此其副詞形式personally可以表達「親自（做某事）」，另外也可以用替換方式in person表示。

④ **Thus, / Therefore, / Hence, / As a result,**

→表達「因此」的用法除了常見的so以外，thus、therefore、hence、as a result等也是常用的轉接詞。這些轉接詞和so一樣，通常較常用於句首。

⑤ **He is so reliable that you don't need to worry about anything.**

→so...that...句型用來強調「如此……，以至於……」，是英語中的常用句型。用在這裡表示「羅賓森先生如此可靠，以至於你不用擔心（他代替出席說明會）」。

單字片語急救包

♥ **attend** *v.* 出席
♥ **represent** *v.* 代表；作為……的代表
♥ **reliable** *adj.* 可靠的
♥ **be on a business trip** 出公差

馬上來練習吧！

想要表示會議延遲時該怎麼說呢？

1. 開頭稱謂

Dear Mr. Lou,

2. 問候句

This letter serves to ____________________.

The conference has been rescheduled ________
____
____
__________________________.

3. 信件主要內容

We apologize ___________________________
____
____
________________________________.

4. 結尾問候句

Thank you for your understanding.

5. 署名

Sincerely,

Jeff

參考解答 Answer

Dear Mr. Lou,

This letter serves to inform you that the conference we set up on September 12 will be postponed to September 19 because of the typhoon. The conference has been rescheduled on September 19 at 19:00 in the Third Conference Room in our office building.

We apologize for any inconvenience that the postponement may have caused.

Thank you for your understanding.

Sincerely,

Jeff

中譯

樓先生您好：

這封信目的是要通知您原訂九月十二日的會議將因為颱風的緣故，將延後至九月十九日。這場會議將重新安排在九月十九日，下午七點，在本公司辦公大樓的第三會議室舉行。

此次延期若造成任何不便，我們在此致歉。

感謝您的體諒。

傑夫 謹上

句型這樣替換也可以

1. **I am sorry to inform you that...** 很抱歉通知您……
2. **The seminar will take place at...** 說明會將在……舉行。
3. **I apologize for...** 在此 ……致歉。

Unit 04 詢問行程異動

寄信時可以這麼說

Dear Nick,

It is said that ① there is some problem ② with the **location** of the conference we arranged for next Monday, April 15 at 10:00 a.m ③. I heard that the original site is temporarily closed down for emergent renovations.So I am writing to **confirm** whether the conference will still be held as planned. If there is a postponement or any change of the conference ④, please contact me at your earliest possible convenience.

Cordially,

Fran Peterson

親愛的尼克：

聽說①我們原先安排在下週一，也就是四月十五日上午十時③的會議**地點**有問題②。我聽說本來的地點因為緊急裝修緣故被暫時關閉。所以我來信是想向你**確認**會議是否仍如期舉行？若是會議有延遲或是任何異動④，請您盡可能在一有空就與我連絡。

法蘭．皮特森 **謹上**

換句話說

① I have heard 我聽說／I was told 我被告知

→郵件中的句子It is said that是常用法，表示「聽說……」，同樣用法還有I have heard that和I was told that。值得注意的是，這種「訊息來源不明或是不重要」的表達方式，通常都是用被動式，像是said和told就屬被動式。

② issue / matter

→同樣都是「問題」，problem、issue和matter也有不同的使用範圍。除了表示「問題」，issue一詞的另外解釋為「議題」，通常表示正式、嚴肅的主題，例如：environmental issue（環境議題）、personal issue（個人問題）；而matter多指「事件」、「事情」，最常的使用方式有：What's the matter、a matter of great concern to the general public等。

③ the President / the new project

→這裡的arrange for後面，除了可以放時間之外，也能放「人」或者「目的」。若是前者，句意為「為某人安排的（會議）」；若是後者，則表示「為某人安排的（會議）」。

④ If the conference has been postponed or changed

→郵件中的原句是用there is...句型表示，這裡也可以將主題conference往前放，替換成if the conference has been postponed or changed...。

♥ **location** *n.* 地點
♥ **confirm** *v.* 證實；確認
♥ **cordially** *adv.* 誠摯地
♥ **It is said that** 聽說

回覆時可以這麼說

Dear Fran,

I feel sorry to tell you that the conference will surely be **put off** **as far as I know** ① at present ②. As you may have heard, the conference room we booked ③ has been **cancelled** without early **notification**. We are now trying to find another place ④ for the conference, and will call you to check your schedule as soon as the location problem is **solved** ⑤.

Again, please accept my apology for all the trouble.

Regards,

Nick Wales

親愛的法蘭：

我很抱歉要告訴你，目前②**就我所知**①，會議將確定會被**延後**。如同你可能已經聽到的，我們預訂③的會議室在沒有預先**告知**的情況下被**取消**了。我們正試著尋找另一個④開會的地點，並會在地點問題**解決**⑤之後，立刻打電話確定您的行程。

謹再次對所造成的困擾致上歉意。

尼克．威爾斯 謹上

換句話說

① according to my knowledge

→according一詞本意為「根據；依據」，常常用來表示資訊、消息來源，所以according to my knowledge表示「根據我的知識」，也就和郵件中的as far as I know同義，表「據我所知」之意。

② presently / currently

→表達「目前」的用法除了郵件中的at present之外，也能直接用副詞單詞presently和currently表示。

③ reserve

→生活中，要預約餐廳、飯店、任何服務，或會議室等，最常使用的動詞就是book和reserve。因此，郵件中的book若改成reserve也是可以的。不過，在有些情況下，兩個動詞分別有常用的搭配方法，例如：book a flight（訂機票）、reserve a table（訂餐廳）。

④ a substitute 替代（品）

→another sth表示「另一個……」，也可以用substitute直接表示「替代（的地點）」。值得注意的是，另一個詞replacement雖然也是「替代」，但其表示的意義為「更換的動作本身」，例如：the replacement of computer equipment（電腦設備的更換）。

⑤ settled

→和problem一詞最常同時出現的動詞為solve，而與其同意義的詞還有settle，同樣表示「解決；確認；安排」。

單字片語急救包

♥ **cancel** *v.* 取消
♥ **notification** *n.* 通知
♥ **solve** *v.* 解決
♥ **put off** 延後
♥ **as far as I know** 就我所知

馬上來練習吧！

想要詢問行程異動時該怎麼說呢？

1. 開頭稱謂　Dear Jeff,

2. 問候句　Thank you for your letter ____________________
____
____.

3. 信件主要內容　Unfortunately, ____________________,
____________________ ____________________.
So I will ____________________________________
____,
____
_____________________________________.

4. 結尾問候句　With best regards,

5. 署名　Kyle Lou

參考解答 Answer

Dear Jeff,

Thank you for your letter of September 9. Unfortunately, I am afraid that I will be unable to attend the conference on September 19 because of a prior arrangement. So I will have our Assistant Manager, Albert Lai, represent me that day.

With best regards,

Kyle Lou

中譯

傑夫：

謝謝您九月九日的來信。很遺憾的是，九月十九日那天因為已經事先安排事情，我恐怕無法出席會議。所以我會派我們的協理亞伯特・賴代表我出席。

凱爾・樓 謹上

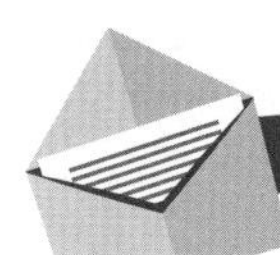

句型這樣替換也可以

1. **I'm glad to notify you that I'll be present in ...**
   很高興通知您我會出席……
2. **It's an honor to have this chance to take part in...**
   有機會參與……是我的榮幸
3. **Please don't feel sorry about...** 請不要對……感到抱歉

Unit 05 無法出席表達歉意

寄信時可以這麼說

Dear Wendy,

Thank you for **inviting** me **to** take part in ① the **celebration** of your 20th Opening Anniversary ②. Unfortunately ③, I will be unable ④ to attend ⑤ it because I have some prior **engagement** on the date of the party. But if there were time left, I would try to make it to the party to send my regards personally in honor of our long-time harmonious collaboration.

Please accept my sincere congratulations! I wish you continued success.

Sincerely,

John Benson

親愛的溫蒂：

謝謝您**邀請**我參加①貴公司二十週年開業紀念**慶祝會**②。令人遺憾的是③，由於我那天已經事先有安排**事情**， 所以將無法④出席⑤參與這場派對。不過如果有多餘時間的話，我會試著趕去派對親自送上我的祝福，向我們長期和諧的合作關係致敬。

謹獻上我誠摯的祝賀！祝您越來越成功！

約翰・班森 謹上

換句話說

① **join in**

→表示「參加；加入」最常用的兩個片語為take part in和join in，兩者可以互相替換。

② **your wedding ceremony 您的結婚典禮／the special showing of the new lines of household appliances 新上市的家電用品系列特別發表會／the 30th Anniversary Celebration of the Founding of your company 貴公司創業三十週年慶**

→這裡的出席場合可以視真實情境作替換。

③ **Sadly / Regrettably**

→表示「可惜地；不幸地」，通常會在句首使用副詞，常見的有：unfortunately、sadly，或regrettably等。

④ **It is impossible for me**

→表示「無法……」除了可以直接說I am unable之外，也可以用「虛主詞it」表示，使用替代句It is impossible for me to attend...，說明「要出席……是不可能的」。

⑤ **show up 露臉**

→表示「出席某個場合」實了attend以外，也可以用show up片語，表示「在……露面」。其他用法還有：Please don't show up late.（請不要遲到）、He didn't show up until 10 o'clock.（他到十點才出現）等。

單字片語急救包

♥ **celebration** *n.* 慶祝會
♥ **engagement** *n.* 約會；安排
♥ **invite sb. to sth.** 邀請某人到某地

回覆時可以這麼說

Dear John,

I feel regretful to hear that ① you are not available for ② our function ③. You are one of our most important **guests** ④; the celebration will not be quite as good without your **attendance** ⑤.

Hope we could arrange a time to **get together** after the celebration.

Wish you all the best.

Yours,

Wendy Cheng

親愛的約翰：

很遺憾知道①你無法參加②我們的盛會③。你是我們最重要的**賓客**之一④；少了你的**出席**，這慶祝會將會遜色許多⑤。

希望在慶祝會之後，我們能安排一個時間**聚聚**。

祝您萬事如意。

溫蒂・程 謹上

換句話說

① It's a pity / shame 好可惜

→郵件中的I feel regretful to...是比較書信的用法，若使用it's a pity或it's a shame代替作為開頭，在語氣上來講，較為親近、口語化。

② can't come by 不能順道拜訪

→come by一詞表示「順道拜訪；短暫露臉」，使用於此處有謙虛之感，表示「他人的時間可貴，只需順道拜訪（就很感謝了）」。

③ annual banquet 年終聚餐／year-end party 尾牙餐會／Opening Ceremony 開幕典禮

→這裡的出席場合可以視真實情境作替換。

④ You made a lot of contribution for our company 你為本公司貢獻許多

→contribution一詞表「貢獻」，使用方式為made contribution to sb./sth.。其動詞形式contribute的使用方式為contribute to sb. /sth.。

⑤ will lose much fun without you

→郵件中，此處欲表達對方對此場合的重要性，因此說明「少了您的出席，這個慶祝會將遜色很多」。除了原句的表示方法以外，也可以用替代句the celebration will lose much fun without you.。

♥ **guest** *n.* 賓客；來賓
♥ **attendance** *n.* 出席
♥ **get together** 聚會；相聚

馬上來練習吧！

想要為無法出席表達歉意時該怎麼說呢？

1.	開頭稱謂	Dear Pamela,
2.	問候句	I am writing to apologize for ______ ______ ______.
3.	信件主要內容	______, ______. Please accept my sincere apology for ______ ______ ______ ______.
4.	結尾問候句	Best regards,
5.	署名	Joseph

參考解答 Answer

Dear Pamela,

I am writing to apologize for not being able to attend the conference held on April 18. Something serious happened to our factory buildings in Kaohsiung the day before the conference, so I had to hurry there at the same night to resolve it. Please accept my sincere apology for not being able to inform you earlier under the circumstances.

Best regards,

Joseph

中譯

親愛的潘蜜拉：

我是寫來對沒有出席四月十八日舉行的會議這件事道歉的。我們的高雄廠房在會議前一天發生了件嚴重的事情，以致於我必須連夜趕到那裡解決問題。在這種情況下，沒能早點通知您，請接受我誠摯的歉意。

喬瑟夫 謹上

句型這樣替換也可以

1. **Thank you for inviting me to...** 謝謝您邀請我參加……
2. **I will be unable to attend...** 我將無法出席……
3. **Please accept my sincere...** 請接受我誠摯的……

Unit 01 洽談建立業務關係

寄信時可以這麼說

Dear Mr. Washington,

I am Jessica Simpson from Luxury Kitchenware Company ①. I have **learned from** the internet ② that your company **is looking for** a kitchenware **manufacturer** to cooperate with ③. I would like to **enquire** whether there's an opportunity for us to establish business relations with ④ you. Our company is known for ⑤ **various** choices of kitchenware, including all kinds of patterns, sizes, and functions. Choose us for your partner will **definitely** be the right choice.

I will be expecting your favorable reply.

Thank you very much.

Sincerely,

Jessica Simpson

親愛的華盛頓先生：

我是奢華廚具公司的潔西卡．辛普森①。我**從**網路②上**得知**貴公司正在**尋求**可以合作③的廚具**製造商**。我想**詢問**一下我們是否有機會能與您建立業務關係④。本公司以**多樣化的**廚具著名⑤，包括各式各樣的圖樣、尺寸、功能。選擇本公司為貴公司的合作夥伴**絕對**會是對的選擇。

期待您肯定的答覆。

非常感謝您。

潔西卡．辛普森 敬上

換句話說

① **This is Jessica Simpson, writing on behalf of Luxury Kitchenware Company.**
我是潔西卡・辛普森，以奢華廚具公司的名義來信。
→除了信中的 "I am XXX from..." 此用法以外，要表明自己的身分也可以用 "on behalf of"（代表……）的片語。

② **your advertisement／your message on the internet**
貴公司的廣告／貴公司公佈在網路上的消息
→這裡可以替換的詞有很多，只要把消息來源替換放入此即可。不過，因為通常公司的廣告資訊都會公佈在網路上，所以這裡用這兩種方式呈現。

③ **to collaborate with**
→ cooperate 和 collaborate 皆可以用 to collaborate/ cooperate with 的片語來表達「和……合作」之意。若想要以名詞形式 cooperation 或 collaboration 來表示，也可以用 be in cooperation/ collaboration with sb. 此片語。

④ **create a partnership with**
→這兩個句型是比較正式的用法，提到的「夥伴」通常指的是商業上的關係，如果要用比較口語的方式，可以說 work together即可。

⑤ **to be famous for sth.**
→to be known/ famous for sth.這個片語除了用在商業溝通之外，也常常用於觀光景點的描述。例如：Taiwan is famous for its beautiful mountains and the hospitality of its people. 台灣以美麗的群山和居民的熱情好客著稱。

♥ **manufacturer** *n.* 製造商
♥ **enquire** *v.* 詢問；打聽
♥ **various** *adj.* 多種的
♥ **definitely** *adv.* 絕對地；毫無疑問地
♥ **learn from...** 從……得知
♥ **be looking for** 尋找；尋求

回覆時可以這麼說

Dear Ms. Simpson,

Thank you very much for your letter. It is our pleasure to ① get your reply to our advertisement online.

Would you kindly **provide** us **with** your latest **catalogue** along with ② a **price list** so that we could have further understanding about your products? ③ **In this case**, we can figure out ④ whether it is best for us to choose you as our partner.

Looking forward to hearing from you soon. ⑤

Regards,

Gary Washington

親愛的辛普森小姐：

非常感謝您的來函。能夠得到貴公司對於本公司在網路上公佈廣告的回覆，是本公司的莫大榮幸①。

能否請您**提供**貴公司最新的**產品目錄**，並附上②**價目表**，好讓我們對貴公司的產品有更進一步的認識呢？③**這樣一來**，本公司便能清楚④貴公司是否會是我們合夥的最佳選擇。

希望很快能得到您的回覆。⑤

蓋瑞・華盛頓 敬上

換句話說

① **We are honored to...**

→在商業的書信中，常常會用到pleasure或honor兩個字的變化，來提高對方的地位，以示自己對對方的重視，並強調隆重感。

② **as well as**

→along with 通常是指「（某物）附帶的某物」，as well as 前後所連接的詞則沒有附屬關係，重要性是同等的。雖然這兩個片語的意思有點差距，但是這邊的用法所表達的意義是一樣的。

③ **Would you kindly provide us with your latest catalogue along with a price list? By this, we could have further understanding about your products.**

→信中的句型是用so that作為連接詞，承接前面主要子句，並直接在同一句中表示「目的」。這邊也可以改用By this代替前面所講的 “Would you kindly provide us with your latest catalogue along with a price list”，再另起一句，表達後面提到的「目的」。

④ **to make sure 確定；確保**

→figure out（弄明白）和make sure兩個片語單獨的意思其實不完全一樣，不過用在這裡整體的意思是差不多的。make sure還常用在「確定某事（不）會發生」，例如：Make sure you turn off the lights before you leave. 離開前請確定你有關燈。

⑤ **We hope to hear from you soon.**

→look forward to本身有期待某事的意思，所以可以用hope to替換。這種期望的口氣，也可以表示對對方回覆的重視。

單字片語急救包

♥ **catalogue** *n.* 目錄；型錄
♥ **price list** *n.* 價目表
♥ **provide...with...** 提供某物給某人
♥ **in this case** 在這樣的情況下；如此一來

馬上來練習吧！

想要建立業務關係時該怎麼說呢？

1. 開頭稱謂

Dear Mr. Chao,

____.

2. 問候句

We have learned that ________________________

____

____

______________.

3. 信件主要內容

____

____

____

____

4. 結尾問候句

We will be glad to be of service to you.

5. 署名

Sincerely yours,

APM Co.

參考解答 Answer

Dear Mr. Chao,

We have learned that you are looking for a car components manufacturer to cooperate with.

We have an excellent reputation for our high-quality car components in the industry for decades, and we are sure that our products will meet all your requirements.

We will be glad to be of service to you.

Sincerely yours,

APM Co.

中譯

趙先生：

我們得知貴公司正在尋找合作的汽車零件製造商。

我們數十年來因生產高品質的汽車零件而享譽業界，我們相信我們的產品一定能符合貴公司所有的要求。

我們將很樂意為您服務。

APM 公司 敬上

句型這樣替換也可以

1. **I have learned from the internet that...** 我從網路上得知……
2. **I would like to enquire whether...** 我想詢問一下是否……
3. **We are confident that...** 我們有信心……

Unit 02 | 詢問**合作意願**

寄信時可以這麼說

Mr. Barney,

I am Angela Hemingway from J.K.B Company. We are one of the **major celebrated commercial agents** of household appliances in Taiwan. ① **To prove** our **incomparable** business performance, the list of awards we've **been granted** in the past 5 years is attached. ②

If you are interested in ③ developing your consumer markets in Asia, we would appreciate an opportunity to cooperate with you. ④

An early reply will be obliged. ⑤

Yours sincerely,

Angela Hemingway

巴尼先生：

我是J.K.B公司的安吉拉·海明威。我們是台灣**最大、最著名的**家電用品**代理公司**之一。① 為了向您**證明**本公司**卓越的**表現，隨信附件是本公司近五年**獲頒**的所有獎項。 ②

如果您有意願③拓展亞洲的消費市場，希望我們能有機會與您合作。④

期待⑤您的回覆。

安吉拉·海明威 敬上

換句話說

① **We stand out from other commercial agents of household appliances in Taiwan.**
本公司在其他家電用品代理公司中出類拔萃。
→stand out (from...)作動詞用，可以用來說明在一群同類事物中，表現突出的佼佼者。相同用法還有top這個字。

② **..., there is a list of awards we've been granted in the past 5 years in the attachment.**
這裡有本公司近五年獲頒獎項的隨信附件。
→attach的原意為「附著；附上」，也可以衍伸作動詞，意思是「把……作為電子郵件的附件」，而attachment則是名詞用法。

③ **wish to V. 希望……**
→If you wish to V.是廣告中常出現的開場白。傳達「你如果想……」，和be interested in一樣有吸引合作目標對象的功能。

④ **It would be our pleasure to cooperate with you.**
能和貴公司合作會是本公司的榮幸。
→在商業的書信中，常常會用到It is our pleasure to...，來提高對方的地位，以示自己對對方的重視，並強調隆重感。

⑤ **be appreciated**
→be appreciated和be obliged的意思大同小異，都有感激、感謝之意，能表達對對方回覆的重視。

- ♥ **major** *adj.* 主要的
- ♥ **celebrated** *adj.* 著名的
- ♥ **commercial agent** *n.* 代理商
- ♥ **incomparable** *adj.* 無雙的；無與倫比的
- ♥ **to prove...** 以證實……
- ♥ **be granted** 授予

回覆時可以這麼說

Ms. Hemingway,

We are very **interested in** collaborating with you, and hope that our products can **win acceptance of** the consumers in Asia. ①

Please offer us a **proposal specifying** ② the terms and conditions of the **cooperation**, and ③ we can discuss the details afterward. ④

Your early reply will be greatly appreciated. ⑤

Sincerely,

Louis Barney

海明威小姐：

我們很**有興趣**與貴方合作，並希望我們的產品能**得到**亞洲消費者**的認同**。①

請提供我們一份**詳述**②**合作**條件的**提案書**，然後③我們再來討論細節。④

期待您的回覆。⑤

路易士・巴尼 敬上

換句話說

① **...gain popularity in Asia market. 在亞洲市場受歡迎。**
→原文用win acceptance of...屬於較消極的用法，表示能「獲得……的認同」；相較之下，用gain popularity in...表示並非只「獲得認同」，而是能「受到……歡迎」。兩者在意義上有些微的差異。

② **detailing 詳細描述**
→detail（細節）是常用的名詞，而它其實也有動詞形式detailing，顧名思義就是詳細描述之意。

③ **so that...**
→信中的and屬對等連接詞，所連接前後的兩個子句是對等的關係。而若替換成so that，整體的意思與原文相同，但從文法層面而言，so that帶出的意思則是「若是……就能……」，表示之前所接的主要子句之目的，有強調「目的」的語意。

④ **we can have further discussion about the details.**
我們可以對於細節有更深入的討論。
→原文和替換句的主要差別在於discuss的詞性變化，以及afterward和further的用法。afterward本身有「以後」的意思，而further則表示「更遠的；更深入的」，若要用further替換句子，則將動詞discuss轉變為名詞形式discussion即可。

⑤ **Please reply as soon as possible, thank you.**
請盡快回覆，謝謝。
→一般商業書信的往來，寄信者都會希望能盡快得到對方的回覆，以便能趕快配合後續事項，所以這一類的句子很常見。不過，通常像原文的說法會較常見，因為在語氣上屬於較禮貌的口吻；而as soon as possible則較強調急迫性之感。

單字片語急救包

- ♥ **proposal** *n.* 提案；求婚
- ♥ **specify** *v.* 具體描述；明確指出
- ♥ **cooperation** *n.* 合作
- ♥ **interested in** 對……有興趣
- ♥ **to win acceptance of...** 獲得……的認同

馬上來練習吧！

收到詢問合作意願的信時該怎麼回覆呢？

1. 開頭稱謂

Dear Sir,

______________________________.

2. 問候句

Many thanks ______________________________

____

__________________.

As you know, we______________________________

______________________________.

3. 信件主要內容

We do hope ______________________________, so

____

____

4. 結尾問候句

With thanks and regards,

5. 署名

Andy Chao

參考解答 Answer

Dear Sir,

Many thanks for your letter of December 5. As you know, we are looking for a partner who can supply us with car components of premium quality.

We do hope we can cooperate with you in the future, so please let us have all necessary information regarding your products. We will inform you of our decision as soon as possible.

With thanks and regards,

Andy Chao

中譯

先生您好：

十分感謝您十二月五日的來函。如您所知，我們正在尋找一個可以提供我們優質汽車零件的合作夥伴。

我們真心希望未來能夠與您合作，所以請您將有關貴公司產品的所有必要資訊予以賜之。一旦我們有所決定就會立刻通知您。

謝謝，並祝福您。

趙安迪 敬上

句型這樣替換也可以

1. **Thank you for your e-mail of...** 感謝您……的郵件
2. **Would you kindly provide us with...** 能否請您提供……
3. **I would be expecting your...** 期待您的……

Unit 03 確認產品細節

寄信時可以這麼說

Dear Mr. Spencer,

I am writing to **follow up on** ① our conversation regarding ② your order for our computer component parts.

Please let us know ③ the exact **quantity** of each item you require and the exact date you'd like us to **deliver** your **goods**, so that we can meet your needs ④ **as much as** we can.

A **prompt** reply by fax ⑤ will be most appreciated.

Sincerely yours,

Mandy Lance

史賓瑟先生：

延續①之前的談話，有些關於②您訂購我們的電腦零組件的事情想在此向您提出詢問。請提供我們③您所需要的每項物品的確實**數量**，以及您希望我們**送貨**的確切日期，以便我們能**盡可能地**配合④您。

麻煩您**盡快**以傳真⑤回覆我們，感激不盡。

曼蒂．朗斯 敬上

換句話說

① **continue on... 繼續……**

→follow up on...的意思為「延續……進行」，口語上來說，其實就是「繼續」的意思，所以也可用continue on作替換，但口氣上來說，是較不正式的用法。

② **as regards / in (with) regard to / concerning**

→regard一詞的意思為「考慮；關於」，其變化的方式有很多，上述的as regards和in (with) regard to都表相同意義。而concerning也是書郵件中常出現的用法，這些詞都會比about更正式。

③ **to notify / inform us... 告知我們……**

→郵件中使用的let sb. know也很常出現在口語使用中，用notify或是inform都會更顯得正式一點。inform的用法與原文中的一樣，而notify的句型則為「notify sb. with sth.」，所以原文可改成 “Please notify us with the exact quantity of...” 。

④ **to satisfy your needs / requirements / conditions**
配合您的需求／要求／條件

→meet one's need(s) 是固定的常用法，表示「某人配合對方的需求」，通常會出現在服務對象的對話中。而satisfy則隱含想「進一步使對方滿意」的積極態度。

⑤ **via fax**

→相較於郵件中用的by，這裡替換使用的via是較正式的用法，有「經由；通過」之意，最常用來表示訊息或資訊的來源或方式。

單字片語急救包

- ♥ **quantity** *n.* 數量
- ♥ **deliver** *v.* 運送
- ♥ **goods** *n.* 商品；貨品
- ♥ **prompt** *adv.* 及時的；迅速的
- ♥ **follow up on** 延續……進行
- ♥ **as much as** 盡可能地

回覆時可以這麼說

Ms. Lance,

I have faxed you the **relevant** ordering **documents**. Please **make sure** that you got it, and take a moment ① to **go through** it.

As for ② the **delivery date**, will it be possible for you to have our goods shipped by December 12? ③

We will **keep you informed** regarding additional orders we may need. ④

Thank you very much.

Your truly,

Neil Spencer

朗斯小姐：

我已經將訂貨的**相關文件**傳真給您了。請**確認**您已經收到了，並且花些時間①**檢閱**。

至於②**交貨時間**，不知道您是否可以在十二月十二日之前將我們的貨品寄出③。

如果我們需要追加訂單，會再**通知您**。④

非常感謝您。

尼爾．史賓瑟 敬上

換句話說

① **spend some time**
→相較之下，take a moment可能會比spend some time更加禮貌，因為moment有「片刻」的意思，傳達出「只耽誤您一些些時間」因為對方的時間很寶貴，不過在意義上，spend some time也是一樣的。

② **in the aspect of... 在……方面**
→as for和in the aspect of都有「關於」的意思，帶有轉折語氣，表示重新提及的話題。這裡可以注意的是，雖然常用的about也有相同的意義，但是缺少轉折口氣，所以較不適合放在句首。

③ **... is it possible for you to have our goods shipped by December 12?**
→兩句在意義上是一樣的，只是時態上的變化有差異。可以注意的是，兩句皆使用了「虛主詞」"it" 的用法，代替後面提到的 "to have our goods shipped by December 12"。

④ **We will renew the number of orders if needed.**
如果有需要，我們會更新訂單數。
→renew本身的意思是「更新」，所以用在這個情境中所表達的意義和原文中的較不同。信件中有較明確的告知對方，若有追加會「主動」告知；而替換句則著重在「訂單數量更新」的可能性。

- ♥ **relevant** *adj.* 相關的
- ♥ **document** *n.* 文件
- ♥ **delivery date** *n.* 交貨日
- ♥ **make sure** 確認
- ♥ **go through...** 仔細檢查……
- ♥ **keep you informed** 通知某人

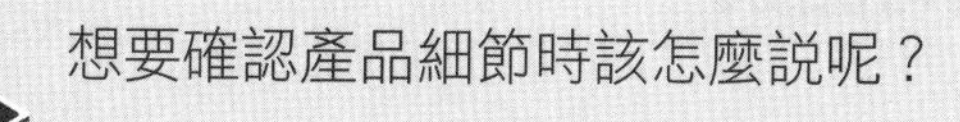

馬上來練習吧！

想要確認產品細節時該怎麼說呢？

1. 開頭稱謂　Dear Gordon,

2. 問候句　I am writing to ______________________________
____
______________________________.

3. 信件主要內容　As you may know, ______________________________,
____
____
____
____
______________________________.

4. 結尾問候句　I appreciate your immediate reply.

5. 署名　Regards,
Lillian Liu

參考解答 Answer

Dear Gordon,

I am writing to inquire how long it will take you to finish this project if we decide to let you do this job. As you may know, we are in urgent need of the goods, so I'm wondering if you can finish the whole process within the next two months.

I appreciate your immediate reply.

Regards,

Lillian Liu

中譯

葛登：

我想詢問，如果我們決定讓貴公司負責這項工作，你們要多久時間才能完成這個案子？你也知道，我們急著要這批貨，所以我想知道你們能不能在兩個月之內完成整個流程？

若能立刻回覆，將感激不盡。

莉莉安・劉 謹上

句型這樣替換也可以

1. **Please let us know the exact quantity of...**
   請告訴我們……確切數量
2. **A prompt reply will be appreciated...** 盡快回覆將感激不盡
3. **I have faxed you the relevant...** 我們已將……傳真給您了

Unit 04 確認合約內容

寄信時可以這麼說

Ms. Nelson,

I am writing this e-mail to enquire if you have **reviewed** our contract and found everything alright. ①

Please inform us if there are any terms that you **object to** ② in the contract, and detail the **preference** you would like us to **adjust** ③ ; otherwise ④ , please send us a copy of the **signed** contract so that we can start our work immediately. ⑤

I hope to hear from you soon.

Thank you very much.

Sincerely,

Leslie Lai

尼爾森女士：

不知道您是否已經**細讀過**我們的合約，並確認沒問題了呢？①

如果合約中有任何您**不同意**②的條件，請告知我們，並且詳細說明您**希望**我們能**變更**的細節③。否則④，煩請立即將**簽好的**合約寄給我們，好讓我們立即⑤開始進行工作。

希望能盡速得到您的回覆。

非常感謝您。

雷絲莉・賴 敬上

換句話說

① **This e-mail is to enquire if you have reviewed our contract and found everything alright. 這封信的目的是為了向您確認是否已經檢閱過我們的合約，並確認沒問題了。**
→郵件中的表達方式是強調「寄信者本人」寫信的動機，而替換的句子沒有寫出I am...，而是以this e-mail作為主詞，強調「信件本身」要達成的目的，兩種方式的切入點不同，但在意義上來說沒有太大的差別。

② **disagree with...**
→相較於郵件中使用的object to，disagree with在使用上較為頻繁，它在正式及非正式的場合都能使用，有「反對；不同意」之意。

③ **, and detail how you like us to adjust it...**
並詳細說明您希望我們怎麼變更細節。

④ **if not,... 否則**
→郵件中這裡的語意是「如果（合約沒問題）就（寄給我們）」，otherwise和if not都有轉折的語氣，可以代替前面提到的子句的反面意思，也就是說，前面提到「如果合約有問題」，後面連接代替的就是「如果合約沒有問題」。這裡可以注意的是，雖然or也有「否則」的含意，但是在口氣上略嫌隨便，所以通常不會在這使用。

⑤ **right / straight away**
→right / straight away 和immediately同義，都表示「立即、立刻」之意，會比口語上常用的soon或是directly等副詞還更正式一點。

♥ **review** *v.* 再檢閱
♥ **preference** *n.* 偏好；希望
♥ **adjust** *v.* 調整；修改
♥ **signed** *adj.* 已簽核的
♥ **object to** 反對；不同意

回覆時可以這麼說

Dear Leslie,

I am sorry to have **kept you waiting**. We have **perused** the contract and we give our approval for ① all the terms. We are very lucky and **grateful** to have such a cooperative partner as you are ② . We will send you a copy of the signed contract by tomorrow ③ , August 11, **at the latest**.

May we have a **pleasant** cooperation. ④

Thank you very such.

Sincerely,

Rose Nelson

雷絲莉：

很抱歉**讓您久等**。我們已經**仔細讀過**合約，並同意①所有條件項目。我們幸運且**很高興**能有像你們一樣，配合度高的合作夥伴。②我們會在明天，也就是八月十一日**之前**③，將簽署好的合約寄送給您。

願我們能合作**愉快**。④

非常謝謝您。

蘿絲・尼爾森 敬上

換句話說

① **give / grant one's permission to 給予許可**
→approval和permission的意義相近，都是「許可；准許」之意，在商業書郵件中常常用到，通常都會和proposal（提案）或contract（合約）一併出現，表示一項提案或合約的通過。

② **How lucky and grateful are we to have such a cooperative partner as you are!**
→替換句和郵件中原文在語意上是一樣的，不過兩者差別在於語氣。將原句的直述句轉換為驚嘆句，更能強調原本想傳達訊息的程度，也就更能顯現寄信者覺得「幸運和高興」的程度。

③ **by Saturday / by next Monday / by the day after tomorrow**
週六以前／下週一以前／後天以前
→「by + 時間」的句型可以表示「在……某時間之前」，其變化的方式只要將時間替換帶入即可。

④ **May we have a great experience working together.**
→商業書郵件中常常用到類似的句子，通常會放在一封信的最後，表示對於與對方合作的期待，展現寄信者的禮貌和重視。

♥ **peruse** *v.* 瀏覽
♥ **grateful** *adj.* 感謝的
♥ **pleasant** *adj.* 愉快的
♥ **keep sb. waiting** 讓某人等
♥ **at the latest** 最晚；遲至

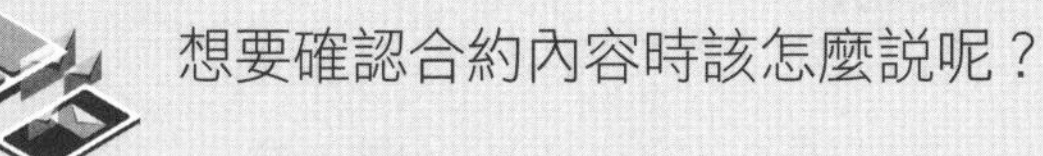

馬上來練習吧！

想要確認合約內容時該怎麼說呢？

1. 開頭稱謂　Dear Catherine,

2. 問候句　I hope you will excuse me for ____.

3. 信件主要內容　We ________________. ________________. Please ____.

4. 結尾問候句　Hope we ____.

5. 署名　Oliver Johnson

參考解答 Answer

Dear Catherine,

I hope you will excuse me for my late reply to your letter dated January 5. We did have gone over the contract and found everything alright. We will sign the contract immediately and have it sent to you by 5:00 p.m. today. Please arrange for production once you receive it and we will remit the down payment tomorrow.

Hope we will have a pleasant cooperation.

Oliver Johnson

中譯

親愛的凱薩琳：

這麼晚才回覆您一月五日的來信，還請見諒。我們的確已經看過合約，一切都沒問題。我們會立刻簽約，並在今天下午五點之前送去給您。請您立刻安排上線生產， 我們明天會將訂金匯過去。

希望我們合作愉快。

奧立佛．強森

句型這樣替換也可以

1. **Please inform us whether you...** 您是否……請告知我們
2. **Would you please send us a copy of...**
   請寄給我們一份……的副本好嗎？
3. **I am sorry to have kept you waiting (for our reply).**
   很抱歉讓您等候（我們的回覆）。

Unit 05 確認服務執行程序

寄信時可以這麼說

Mr. Henderson,

I am Janice White from Wonderful Furniture. I am here to inform you of the detailed **procedure** of the **customized** sofa delivery service you ordered last Saturday.

Since this is a customized product ①, we will need to have the **measurement** of the room where you would like to keep the sofa ②. We would like to ask if this Wednesday night would **be available** for you ③. If so ④, our staff will be there at 7 o'clock sharp ⑤ that day. After we get the numbers, we will **have** your sofa **ready** in two days, and deliver it at the same day, which is by Friday.

Hope to hear from you soon. Thank you.

Sincerely,

Janice White

韓德森先生：

我是美好傢俱的珍妮絲・懷特。這裡要向您說明您在上星期六預定的**客製化**沙發運送服務的執行**程序**。

由於是客製化產品①，我們會需要您沙發放置空間的**丈量數據**。②我們想問您這個星期三是否**方便**③。如果是的話④，我們的員工會在七點整準時⑤到達。有了丈量數據後，我們會在兩天之內製作**完成**您訂製的沙發，並在當天送達您家裡，也就是星期五。

希望您能盡快回覆。謝謝。

珍妮絲・懷特 敬上

換句話說

① due to / owing to / because of the customized service we provide,... 由於我們提供客製化的服務，⋯⋯

→上列的三種片語在意義上是由because衍伸而來，中文一樣都是「因為」的意思，但是because後可以接完整句子，due to / owing to / because of後只能接名詞或名詞片語。

② This is a..., so we will need to...

→原文句型和代替句型都是「因為⋯⋯所以⋯⋯」，不過要注意的是，since / because（因為）和so（所以）兩個連接詞不能同時出現，和中文大不相同，而意義上雖然一樣，前者較著重在原因；後者則是結果。

③ be fine for you

→fine一詞在英語中的用法很廣，除了形容某個事物「很好」之外，當有人問 "which one do you like, tea or coffee?" 這種選擇問句時，回答可以用 "Both are fine for me." 表示兩者都是可接受的。郵件中的情況也是類似的用法，表示「（七點）是可接受的」。

④ if this would be the case,...

→原句的so本身有代替前面提到的內容之功用，所以if so便能表示「如果是（星期三方便）的話」，而if this would be the case也是相同的意思，不過相較之下會比if so更正式一點。

⑤ at exact 7 o'clock

→sharp和exactly都表示「整點」，在英文中常用到，表示「某個整點時間」且「分秒不差」。

- ♥ **procedure** *n.* 程序
- ♥ **customized** *adj.* 客製化的
- ♥ **measurement** *n.* 測量
- ♥ **be available** 可用的；有空的
- ♥ **have sth. ready** 把⋯⋯準備好

回覆時可以這麼說

Ms. White,

Thank you for informing me.

Wednesday at 7 p.m. would be great for me ① . And for the delivery time on Friday, I hope you can come here at a time before 8 o'clock and after 5 o'clock. I will not be at home until 5, and my family and I are going to have a big dinner at 8, so we will not also be at home then ② .

Thank you again for always consider customers as a **priority** ③ . Your service is **efficient** and **absolutely satisfying**, not to mention your attitude. It also would be **undoubted** fact that ④ the products you make **is of** high-quality.

See you on Wednesday.

Sincerely,

Adrian Henderson

懷特女士：

謝謝您的告知。

星期三七點我很方便。①至於星期五的運送時間，我希望您能在五點之後、八點之前到達。五點前我還不會在家；而八點後我會和家人去外面聚餐，所以那時②也不會在家。

再次謝謝您總是以顧客**至上**。③貴公司的服務又**有效率**又**令人相當滿意**。你們的產品想必也**無疑**④是很**有**品質的。

星期三見。

亞德安．韓德森 敬上

換句話說

① **It would be great if you come Wednesday at 7.**
→郵件中的句型算是替代句型的簡略用法，兩者的意思一樣，都是為回應寄件者詢問時間的正面答覆。

② **at that time / by then**
→時間副詞在英文中通常都是放在句尾，和中文大不相同，而表示「當下」的詞也有很多種，at that time和by then都能代替原文中的then。

③ **...value customers 重視消費者**
→對於某人的重視可以用see / consider sb. as a priority（優先事項），或是value （重視）sb.來表達。

④ **Undoubtedly,... 無疑地**
→undoubted一詞表示「沒有疑問的」，如果要加入undoubted到句子中，需要使用「虛主詞」（it），而若用副詞形式undoubtedly，則可以直接至於句首，使句型較精簡。

♥ **priority** *n.* 優先事項
♥ **efficient** *adj.* 有效率的
♥ **absolutely** *adv.* 絕對地；完全地
♥ **satisfying** *adj.* 令人滿意的
♥ **undoubted** *adj.* 無疑的
♥ **be of...** 具有……特性

馬上來練習吧！

想要確認服務執行程序時該怎麼說呢？

1. 開頭稱謂　Dear Customer,

2. 問候句　We are writing to ______________________________
____
____.

3. 信件主要內容　We've noticed that ____________________.
____
______________________.
If ____
____
______________________________.

4. 結尾問候句　Best regards,

5. 署名　Happy House Service

參考解答 *Answer*

Dear Customer,

We are writing to inform you of the procedure of our service.

We've noticed that you placed an order of our Domestic Cleaning Services; however, you haven't finished the payment procedure. We only accept payment in 3days after the day of the order. So please pay the fee as soon as possible so that we can confirm the following service you ordered.

If there's anything we can do to help, please feel free to let us know.

Best regards,

Happy House Service

中譯

親愛的顧客：

此封來信的目的是要向您確認本公司服務的執行流程。

我們注意到您已下訂本公司的到府打掃服務，卻尚未完成支付程序。我們只接受下訂單後三天以內的付款，所以請您盡快繳清費用，我們才能確認之後的服務。

如果有任何有需要我們幫忙的地方，請不吝請教。

開心到府服務公司 敬上

句型這樣替換也可以

1. **We are writing to…** 我們來信是為了……
2. **Place an order of…** 下訂單購買……
3. **You haven't finished…** 您尚未完成……

Unit 01 預算調整

寄信時可以這麼說

Mr. Nelson,

I am writing to inform you a slight change in our **budget**.

Last week, we confirmed our budget in the morning meeting in which everyone agreed ① on spending 30 **percent** of the **gross revenue** on our **advertisement** next month. However ②, the revenue of our company has **decreased** a lot this year, which means we would have to cut ③ our advertisement budget **from** 100 thousand **to** 60 thousand. We would also have to change a few details ④ of our advertisement plan.

Sincerely,

Daniel Chen

尼爾森先生：

我來信是要告知你**預算**調整的事項。

上個星期，我們已經在早會上確定我們的預算，並達成共識①要把公司**總營收**的**百分之**三十花在下個月的**廣告**花費。然而②，公司今年的收入**下降**了不少，代表我們必須將廣告花費**從**10萬減少③**到**6萬。我們也要更改原先廣告的一些細節④。

丹尼爾・陳 敬上

換句話說

① come to a conclusion 以……做結論

→conclusion一詞表示「結論；結尾」，用在此處可以表示「做出（預算金額）的結論」。其他和conclusion相關的用法還有in conclusion，表示「最後」。

② Nonetheless

→此處的however為轉折詞，承接與上句相反的語氣，表示「不過；但是」，相同的用法還有nontheless，通常兩者都至於句首，後面會用逗號隔開下一個句子。

③ limit

→limit除了可以作名詞表「上限；限額」以外，也能作動詞表「限制」，用法和郵件中的cut相同，兩者可以互相替換。

④ make a few adjustment 做稍微的調整

→adjustment一詞表示「調整；適應」，通常習慣的搭配動詞為make，也就是make a few adjustment。

單字片語急救包

- ♥ **budget** *n.* 預算
- ♥ **percent** *n.* 百分比
- ♥ **gross revenue** *n.* 總營收
- ♥ **advertisement** *n.* 廣告
- ♥ **decrease** *v./n.* 減少；下降
- ♥ **from…to…** 從……到……

回覆時可以這麼說

Mr. Chen,

This is such depressing news. ① Our department has had **discussed** the whole plan for a month, and finally finished it yesterday. But now we have to start from the beginning ②.

However, the revenue is always **unpredictable**, so does ③ the budget. We cannot change the **fact** of the cut of the budget ④; all we can do is to **accept** it and **move on**. Our team will start on our new advertisement plan as soon as possible. But we cannot **guarantee to** finish it by this week, for this is really a sudden change for us.

Sincerely,

Jerry Nelson

陳先生：

這真是個令人難過的消息。①我們部門**討論**這整個計畫已有一個月了，終於在昨天結束，但現在我們又要從頭來過②了。

不過，收益是**很難預料的**，預算也是③。我們無法改變要刪減預算的**事實**④，只能**接受**並且**繼續努力**。我們的團隊會盡快開始討論新的廣告計劃，不過我們不能向你**保證**能在這個星期結束，因為這對我們來說是一個很突然的改變。

傑瑞・尼爾森 敬上

換句話說

① We are really depressed by the news.

→depress一詞表示「使憂鬱；使沮喪」，若要使用其動名詞形式depressing，主詞須為「製造出憂鬱的事物」，在此應為the news；而若要使用depressed形式，主詞則為「感到沮喪的人」，在此應為we（our department）。

② start it all over

→表示「重新開始；從頭來過」除了可以用郵件中的start from the beginning（開端）以外，還可以用替代句start it all over。其中的「需主詞it」代表的是前面提到的the whole plan。

③ not to mention... 更不用說是……

→郵件中的so does表示前面提到的the revenue和後面連接的the budget是一樣不可預測的（unpredictable）。若換成替代句，則有強調口吻，表示「更不用說（the budget有多變化多端的了）」。

④ the fact that our budget has to be cut

→the fact的句型主要有兩種，其一為郵件中的「the fact of + N」，其二為「the fact that + 子句」。因此此句可改為the fact that our budget has to be cut。

單字片語急救包

- ♥ **discuss** *v.* 討論
- ♥ **unpredictable** *adj.* 不可預測的；變化多端的
- ♥ **fact** *n.* 事實
- ♥ **accept** *v.* 接受
- ♥ **move on** 接受現實，準備好接受新的體驗
- ♥ **guarantee to** 保證

馬上來練習吧！

想要表示預算調整時該怎麼說呢？

1. 開頭稱謂　Dear Wendy,

2. 問候句　First of all, ______________________________
____
______________________________.

3. 信件主要內容
However, ______________________________.
____
______________________________.
I am wondering if ______________________________
____
______________________________.

4. 結尾問候句　Looking forward to hearing from you soon.

5. 署名　Yours truly,
Michael

參考解答 Answer

Dear Wendy,

First of all, I must say that your proposal on the new project is excellent. However, I have bad news for you. I was told that we have to cut the budget down to ten thousand dollar. I am wondering if it is possible for you to come up with another plan which we are able to safe more money.

Looking forward to hearing from you soon.

Yours truly,

Michael

中譯

親愛的溫蒂：

首先，我必須説你為新企畫寫的提案真的太棒了。不過，我有個壞消息要告訴你。我被告知我們的預算要被刪減為一萬元了。我在想你是否能重新想出一個能讓我們省更多錢的提案。

期待你的答覆。

麥可 敬上

句型這樣替換也可以

1. **I must say that…** 我必須説……
2. **I have bad news for…** 我有個壞消息要告訴……
3. **I am wondering…** 我在想……

Unit 02 行銷策略討論

寄信時可以這麼說

Ms. Watson,

I am writing to discuss about the big sale we are going to hold this weekend.

Since this is our first year of business ①, I think it would be a great chance for us to **widen** our market and **introduce ourselves to** the public by our high-quality products. To achieve this goal, I think a **catchy** slogan as well as attractive **flyers** ② would be needed during our **campaign**. As for the price of our product, how do you think about providing a buy-one-get-one-free ③ deal? By this, I guarantee that our company and products will cause a **sensation** on the market ④.

Please let me know if you agree or not.

Sincerely,

Lucas Moore

華生女士：

我來信的目的是要和您討論這個週末將要舉辦的促銷活動。

因為今年是我們公司創立的第一年①，所以我認為經由我們的高品質產品，這會是一個讓我們**擴展**市場，以及**向**大眾**介紹**公司的好機會。為了達成這個目的，我認為我們的**活動**當天會需要一個**琅琅上口**的口號，還有吸引人的**宣傳單**②。至於價錢的部分，你認為買一送一③折扣如何？這樣的話，我保證公司和我們的產品絕對會在市場上引起很大的**騷動**④。

請告訴我您是否同意。

盧卡斯．摩爾 敬上

換句話說

① **our company is going to release a new brand**
公司將要推出新的品牌
→舉辦促銷的目的有很多，有時是為了宣傳新創公司，有時可能是宣傳新推出的品牌或產品，這時可以用new brand / product release表示。

② **not only a catchy slogan (but) also attractive flyers**
→郵件中的as well as其實和and一樣，都屬對等連接詞，表示「和」。這裡如果將句型改成not only…(but) also則有強調後者的作用，因此這句替換句可譯為「不但是琅琅上口的口號，而且吸引人的宣傳單（會需要）」。

③ **20 percent off 八折**
→常見的促銷方式有買一送一buy one, get one free，和「打X折」，用X percent off表示。值得注意的是，英文的折扣說法和中文不同，若要表示打八折，則要用「扣掉百分之二十」來表示，所以英文為20 percent off。

④ **impress our potential customers 會令潛在客戶印象深刻**
→impress一詞表示「使……印象深刻；給……留下深刻印象」，用法為主動的「impress sb. with sth.」，或被動的「sb. impressed by sth.」。

- ♥ **widen** *v.* 加寬
- ♥ **catchy** *adj.* 琅琅上口的；好記的
- ♥ **flyer** *n.* 宣傳單
- ♥ **campaign** *n.* 活動
- ♥ **sensation** *n.* 知覺；引起騷動的事物
- ♥ **introduce…to…** 向……介紹……

回覆時可以這麼說

Mr. Moore,

In my opinion ①, your idea of slogan and flyer is really **creative**, and would truly attract many customers; however, I think the buy-one-get-one-free deal is a **risky** decision for a **start-up** like us ②. I think a 30 percent discount would be enough.

What's more, I have ③ a new idea of having an official **mascot** represent our company. By this, the image of our company can easily **root in** the customers' mind.

If there are no further questions ④, I will soon inform the Marketing Manager of our proposal, and have him **dispatch** the work down to their department.

Sincerely,

Ava Watson

摩爾先生：

依我看來①，你的口號和宣傳單的想法很**有創意**，也一定能吸引很多消費者。不過，我認為買一送一的折扣對於我們這種**新創公司**來說，**風險太高**了②。我認為打七折的折扣就足夠了。

再來，我還有③一個新的想法，用一個官方的**吉祥物**來代表我們公司。這樣一來，公司的形象就能輕易地**深植在**顧客的腦海裡。

如果沒有其他問題④，我就馬上將我們的提案告知行銷經理，請他**分派**工作給他的部門。

艾娃・華生 敬上

換句話說

① From my point of view

→要表達自己的意見時，可以用In my opinion或From my point of view作為句子開頭，表示「依我看來；從我的觀點而言」。

② a decision of high risk with unpredictable consequences a start-up like us may not afford.

是一個高風險的決定，而其後果我們這種新創公司可能無法承擔。

→risk一詞表示「風險」，可以直接用形容詞risky修飾很有風險的事物，或是用替換句a...of high risk。

③ come up with

→通常表示「想到某想法」時，會使用come up with an idea片語。另外，sth. occur to sb.也是相似用法，但是該注意主詞和受詞的位置是相反的，表示「某想法出現在某人腦中」。

④ If we reach an agreement 如果我們成共識

→通常信件的最後，會向對方確認兩人是否有達成共識，或是否有其他疑問，以便後續的提案執行。所以郵件中的if there are no further questions是常用句，表示「如果沒有其他疑問」，也可以用if we reach an agreement表示「如果我們達成共識的話」。

- ♥ **creative** *adj.* 有創意的；創新的
- ♥ **risky** *adj.* 有風險的
- ♥ **start-up** *n.* 新創公司
- ♥ **mascot** *n.* 吉祥物
- ♥ **dispatch** *v.* 分派；派遣
- ♥ **root in** 植入

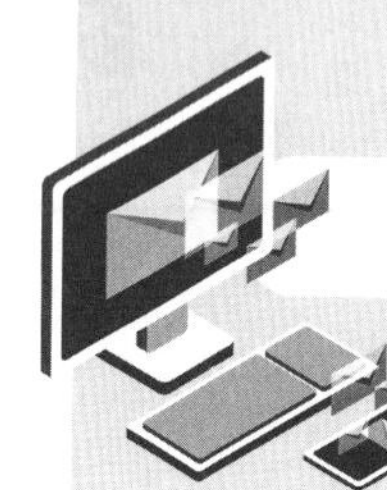

馬上來練習吧！

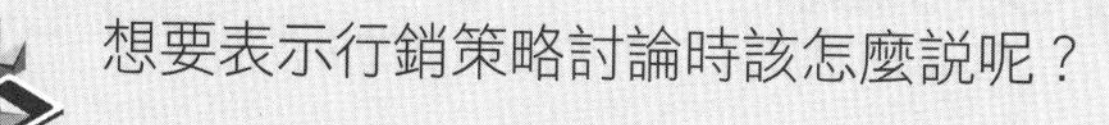

想要表示行銷策略討論時該怎麼說呢？

1. 開頭稱謂　Dear Jake,

2. 問候句　I am writing for the purpose of ______________________

____.

3. 信件主要內容

Since ____.

____.

____

________________________.

Please ____

____

_________________________________.

4. 結尾問候句　Thank you ____.

5. 署名　Amber Kuo

參考解答 Answer

Dear Jake,

I am writing for the purpose of the sale we are going to have next week.

Since this will be the biggest campaign we are having this year, I really hope we can have it perfectly done. We already come up with the products we are going to sell, the location, and the decoration of the event that day. The last step is to fix the prices of every product on the sale. Please think about this with our marketing team and reply me as soon as possible.

Thank you for your hard work.

Amber Kuo

中譯

親愛的傑克：

我來信是為了討論我們下星期要舉辦的大拍賣活動。

因為這會是我們一年中最大的活動，所以我真的希望我們能完美地結束。我們已經想出要販售的產品、地點，和當天要擺地裝飾。最後一步就是要將所以產品都定價。請和我們的行銷團隊討論一下，並且盡快回覆我。

謝謝您的辛勤努力。

安柏・郭

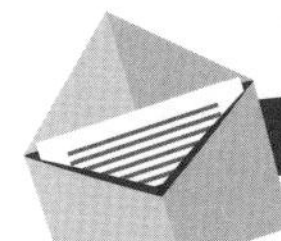

句型這樣替換也可以

1. **Come up with...** 想出……
2. **Think about...** 想一想……
3. **Thank you for...** 謝謝你的……

Unit 03 | 廣告投放

寄信時可以這麼說

Ms. Bailey,

I am writing to discuss the public reaction for our advertisement **released** last week.

According to official **statistics**, we successfully increase ① our **sales volume** and made a **profit** of 30 thousand dollars since the release. Among all the age groups, our products mainly ② attract people of middle age from 35 to 50 years old, men especially ③. To keep improving, we will need to come up with a new advertisement which can both **maintain** the high sales volume among middle-aged men but also interest young people ④. There is still a lot room to improve.

Sincerely,

Marry Lindeman

貝利女士：

我來信是為了向您討論上週廣告**投放**後的大眾反應。

根據官方**數據統計**，自從廣告投放後，我們成功地增加①了**銷售量**，**利潤**也達到三萬元。在所有年齡層之中，我們的產品主要②吸引年齡從三十五到五十歲的中年人，而男性佔多數③。為了更進步，我們會需要想出一個新的廣告，能同時**維持**對中年男子的高銷售量，也能吸引年輕人的目光。我們還有很大的進步空間④。

瑪莉・林德曼 敬上

換句話說

① get a rise on

→表示「上升；增加」除了可以用郵件中的increase以外，還可以使用rise一詞，例如：get a rise on sales volume。

② mostly / primarily / especially

→表示「主要地」除了用郵件中的mainly以外，還有很多同義副詞，例如：mostly、primarily，以及especailly都可以替換。

③ take up a large proportion

→郵件中原句men especially表示「吸引到的中年人之中，又屬男性居多」。除此之外，還可以用take up...proportion表示。take up片語表示「佔據」，proportion表示「部分；比例」，因此take up a large proportion表示「占大部分」。

④ our next goal is to target at young people

我們的下一個目標是要把年輕族群當作主要對象

→target at sb. / sth.表示「把（某人／某事）是為目標」，因此替換句target at young people表示「以年輕人為目標」已足夠表達郵件中原句的意義。

- ♥ **release** *v.* 釋放；公布
- ♥ **statistics** *n.* 數據統計
- ♥ **sales volume** *n.* 銷售量
- ♥ **profit** *n.* 利潤
- ♥ **maintain** *v.* 維持
- ♥ **according to** 根據

回覆時可以這麼說

Ms. Lindeman,

Congratulations! This is how people said "well begun is half done." We deserve this success after all the work and research we had done before ① the advertisement.

Now, our new goal to attract young people would not be difficult since our team is full of young people with **creativity** and **imagination**, our products can certainly be loved by ② youngsters. Along with ③ these two group of people, I think **female** ④ is also a great potential market which we may **benefit from**. Try to include every **customer base** into our target so that we can increase our **market share**.

Sincerely,

Maggie Bailey

林德曼女士：

恭喜！這就是人們說的「好的開始是成功的一半」。我們在廣告之前做了那麼多功課和研究①，這個成功絕對是我們應得的。

現在，我們吸引年輕人的新目標應該不會很困難，因為我們的團隊充滿著有**創意**又有**想像力**的年輕人，我們的產品絕對會受年輕族群喜愛②的。除了③這兩個族群，我認為**女性**④也是一個很大的潛在市場，我們能**從中獲益**。試著將所有**消費族群**納入目標對象，這樣我們就能擴大**市場佔有率**。

瑪姬．貝利 敬上

換句話說

① we spared no effort in 不遺餘力地做／do our utmost 盡最大努力

→effort一詞表示「努力」，因此spare no effort表示「沒有遺留任何努力」也就是「不遺餘力」之意，為固定常用語。另外的do one's utmost也屬慣用句，表示「盡最大努力」。

② popular among

→表示「受……喜愛」可以用「be loved by sb.」或者「be popular among sb.」，因此本句可替代為be popular among youngsters。

③ In addition to

→表示「除此之外，還……」除了可以用郵件中的along with...以外，還可以用in addition to...。值得注意的是，同樣譯為「除了」的expect一詞，在意義上是不同的。其表示「除了……以外」。

④ office worker 上班族／double-paying family 雙薪家庭／one-parent family 單親家庭／Newlywed 新婚夫妻

→消費族群除了可以用年齡、性別區分之外，也有其他種分類方式，例如：上班族office worker、雙薪家庭double-paying family、單親家庭one-parent family，或是新婚夫妻newlywed。

- ♥ **creativity** *n.* 創意
- ♥ **imagination** *n.* 想像力
- ♥ **female** *n.* 女性
- ♥ **customer base** *n.* 消費族群
- ♥ **market share** *n.* 市場占有率
- ♥ **benefit from...** 從……受惠

馬上來練習吧！

想要表示廣告投放時該怎麼說呢？

1. 開頭稱謂　Dear Katrina,

2. 問候句　I am writing to discuss ______________________________ ____.

3. 信件主要內容　According to ______________________________. ____ ______________. ______________, __________________________ ____ ________________________________.

4. 結尾問候句　Please ________________________________ ________________________________.

5. 署名　Sincerely,
Lana Cooper

參考解答 Answer

Dear Katrina,

I am writing to discuss the public reaction for our advertisement released last week.

According to official statistics, our new advertisement didn't get well feedback. We only increase our profit for 5 percent after the release. To improve this, we are going to hold a meeting tomorrow to discuss over this issue.

Please make sure to be there on time.

Sincerely,

Lana Cooper

中譯

親愛的卡翠娜：

我來信是為了討論上週廣告投放後的大眾反應。

根據數據統計，我們的新廣告並沒有得到很好的回應。自從投放後，我們只增加了百分之五的利潤。為了改進這個問題，我們會在明天安排一場會議來討論此事。

請記得準時出席。

拉娜．庫柏 敬上

句型這樣替換也可以

1. **According to official statistics…** 根據官方數據顯示……
2. **We didn't get well feedback on…**
   我們沒有得到很好的回應在……
3. **To improve this…** 為了改善……

Unit 04 | 尋找贊助商

寄信時可以這麼說

Mr. Huang,

This is Jimmy Lin from Happy Reader **Publisher**. We are one of the most book-**awarded** publishers these years. We are now happy to introduce you an **international** book fair we are going to hold on September 20 ①.

In this book fair ②, we are going to **invite** 10 famous writers to **share** their life stories ③ **with** the public and the medias. We would be very grateful if we could accept the care of your company, Share And Like Media, to **sponsor** this event. We will offer you a 60 percent off discount on all of the books we publish in return ④.

Please **think over** this deal. Hope to hear from you soon.

Sincerely,

James Brown

黃先生：

我是樂讀**出版社**的吉米・林。我們是近年來出版過最多**得獎**書籍的出版社之一。今天，我們很高興地向您介紹將在九月二十日舉辦的**國際**書展①。

在這個書展②中，我們會**邀請**十位知名的作家到現場，**向**民眾和媒體**分享**他們的人生故事③。若能承蒙貴公司分享案讚媒體的關照**贊助**，我們將不勝感激。我們會向貴公司提供四折的折扣優惠，作為回報④。

請您好好**考慮**這個提議。希望盡快收到您的回覆。

詹姆士・布朗 敬上

換句話說

① to announce the launch of the new book written by J.K. Rowling
公布J．K．羅琳的新書發表會
→通常一個正式的公開活動會用launch一詞，表「發表會；啟動儀式」，其使用範圍很廣，例如：the launch of new products，還可做動詞用，例如：launch a party。

② exhibition 展覽
→從字面上來看，雖然fair和exhibition兩的詞都做「展覽」解釋，但是fair通常屬於有商業買賣的展覽活動，而exhibition則為單純藝文欣賞類的展覽活動，兩者有些微的差異。所以，此處的「書展」應該使用book fair而非book exhibition。

③ great works 偉大的作品
→通常對於藝術家、作家等的作品，都會用「偉大的」great來形容，因此郵件此處也可以替換用great work，表示知名作家的「偉大作品」。

④ as a return for your favor / in exchange for your favor
作為交換
→return一詞表示「回報；報答」，因此郵件中的in return表示「作為（贊助的）回報」。相似用法還有as a return for sth.或是in exchange for sth.，要注意的是，兩個片語後的受詞皆不可省略。

單字片語急救包

♥ **publisher** *n.* 出版商；發行機構
♥ **award** *v./n.* 授予；獎項
♥ **international** *adj.* 國際的
♥ **invite** *v.* 邀請
♥ **sponsor** *v.* 資助；贊助
♥ **share...with...** 和……分享……
♥ **think over** 考慮

回覆時可以這麼說

Mr. Brown,

After the discussion among our company, we are happy to inform you that we are willing to sponsor ① your event.

Happy Reader Publisher has always been a **reputable** publisher in our nation ②. The events you hold are undoubtedly meaningful. It is our honor to be a part of you and show our support ③. By this international book fair, more and more people will **take notice of** your company and the **importance** of reading ④.

Wish you a successful book fair.

Sincerely,

Denis Huang

布朗先生：

經過本公司的討論，我們很高興告知，我們願意贊助①你們的活動。

樂讀出版社在本國②一直是**享譽盛名的**出版社，貴公司舉辦的活動一定也很有意義。能贊助貴公司並表現支持③也是本公司的榮幸。透過這次的書展，越來越多人會**注意到**你們，以及閱讀的**重要性**④。

祝你們書展成功。

丹尼斯．黃 敬上

換句話說

① **fund 資助**

→sponsor「贊助」一詞表示給予金錢上的幫助，也可以用fund一詞替換，表示「資助；為……提供資金」。而fund一詞也可作名詞用，例如：a pension / trust fund（養老／信託基金）。

② **throughout the country**

→表示「（遍及）全國」除了可以用in our nation以外，也可以用throughout the country表示。Throughout一詞的使用範圍很廣，例如：throughout the day、throughout the contest等，通常可用來表示「一整段時間內……」。

③ **high regard for book-reading 對於閱讀書籍的重視**

→regard for sth.表示「對……的重視」，因此在此可將郵件中的support替換為high regard for book-reading，更能確切的將「支持」具體化。

④ **your company will gain more and more recognition 貴公司將能被更多人認識**

→郵件中的take notice是固定用法，表示「注意到……」。在此，寄信者想表達的是「經過這次的書展，出版社將會更有名」。除了take notice以外，也可以用gain recognition片語，再用more and more修飾「漸增」的狀態。

♥ **reputable** *adj.* 聲譽好的
♥ **importance** *n.* 重要性
♥ **take notice of...** 注意到……

馬上來練習吧！

想要尋找贊助商時該怎麼說呢？

1.	開頭稱謂	Dear Sir,
2.	問候句	This is____________ from_________________.
3.	信件 主要內容	We are one of the most _______________________, ____________________________________. So, we are planning to __________________ ____ _________________________________.
4.	結尾 問候句	We would be very grateful if____________________ _______________________________________. _______________________________________. Hope to hear from you soon.
5.	署名	Sincerely, Tony Franklin

參考解答 Answer

Dear Sir,

This is Tony Franklin from Winner Education.

We are one of the most excellent cram schools these years, and our goal is to help more and more students to fulfill their dream of entering their ideal colleges. So, we are planning to expand our classrooms and upgrade our inner equipments at the end of this year. We would be very grateful if we could accept the care of your company, Better Media, to sponsor our company. We will grant you a Certificate of Appreciation to show our appreciation.

Hope to hear from you soon.

Sincerely,

Tony Fanklin

中譯

先生您好：

我是贏家教育的托尼・富蘭克林。

本公司是近年來最優秀的補習班之一，我們的目標是幫助越來越多的學生完成他們進入理想大學的夢想。因此，我們預定會在今年年底擴展教室，並升級室內設備。若能承蒙貴公司最佳媒體的關照贊助，我們將不勝感激。我們願意授予貴公司一張感謝狀以示我們的感激。

期望您的盡早回覆。

托尼・富蘭克林 敬上

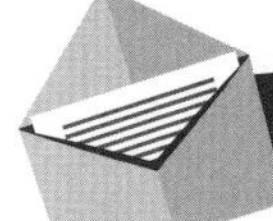

句型這樣替換也可以

1. **We are one of the most excellent…** 我們是最優秀的……
2. **Our goal is to help more and more students to…**
   我們的目標是幫助越來越多學生……
3. **We will be very grateful if…** 如果您……我們會非常感激

Unit 05 進度確認

寄信時可以這麼說

Mr. Lee,

I am writing to confirm the progress of our new housing project.

Yesterday we just settled our deadline of **construction** on December 20, 2020, which means we still have two years to complete from the starting date, December 20, 2018. However, the government just made an unexpected **announcement** this morning and shortens ① our construction duration within one and a half year, so we will have to make sure **to meet the deadline** on June 20, 2020, half a year earlier than ② the **previous** plan. This means we will have to start working on the project sooner, and **work over time** on weekends.

This is a sad new but I believe if we all stick together ③, to complete in time is not a **mission** impossible ④.

Sincerely,

Nick Wang

李先生：

我來信是為了向您確認我們新住宅計畫的進度。

昨天我們剛協議好**建設工作**的截止日期，訂在2020年12月20日，也就是說，從我們的開始日期2018年12月20日起，還有兩年的時間。不過，政府今早**公布**了一個意想不到的消息，把我們的建設時間縮短①為一年半。因此，我們必須要確認能在2020年6月20日**完工**，也就是提早②**原**計畫半年的時間。這代表我們必須提早開工，還要在假日**加班趕工**。

這是一個難過的消息，不過我相信，只要我們同心協力③，準時完工並不是不可能的**任務**④。

尼克・王 敬上

換句話說

① unexpectedly announced to

→announce一詞表「公布；聲明」，有動詞用法「announce sth」和鳴詞用法「make an announcement」。因此此句可改為announced to shorten our construction duration...。

② prior to

→表示「比……提早」除了可以用early的比較及行式earlier to...以外，也可以直接將prior to放在原訂時間之前即可。

③ unite as one

→表示「團結一致」除了可以用郵件中的stick together以外，也能用替代句unite as one，表「結合為一」。

④ the situation would not be as hopeless as we think

情況不會有我們所想的糟

→郵件中此處是要表達安慰，表示「雖然要提早完工，但是我們做得到」之意。所以可以用替換句the situation would not be as hopeless as we think，表示「情況沒有那麼糟」。

單字片語急救包

- ♥ **construction** *n.* 建設；建築物
- ♥ **announcement** *n.* 公告；聲明
- ♥ **previous** *adj.* 先前的
- ♥ **mission** *n.* 任務
- ♥ **to meet the deadline** 趕上截止日期
- ♥ **work over time** 加班

回覆時可以這麼說

Mr. Wang,

This is such a **hasty** decision. But don't worry; I used to ① make good **preparation** for constructions **in advance** just to **avoid** the situation of failing to meet the deadline from happening ②. I suggest we **apply for** more workers from other construction projects with much longer construction duration, so we can make faster progress ③ each day.

By the way, the safety of the workers and residents who will live here in the future is a priority. We don't want to risk people's lives ④ just to get our construction done in time by **cutting corners**.

Sincerely,

Benjamin Lee

王先生：

這真是個**倉促的**決定。不過別擔心，我一向習慣①在建設前**提前**做好**準備**，**避免**趕不上截止日期的情況發生②。我建議從其他有較長建設時間的建設計畫**申請**更多的工人，這樣我們每天的進度就能更快一點③。

順帶一提，工人和未來住戶的生命安全是首要考量。我們不能為了準時完工而**偷工減料**，危及人命④。

班杰明・李 敬上

換句話說

① **always / often / sometimes**

→郵件中的used to是常用片語，表示「習慣」。在此，除了可以用used to以外，也可以替換成頻率副詞修飾後面的動詞，例如：always、often、sometimes等，依序表示為「一直；總是」、「常常」、「有時候」。

② **in case we fail to meet the deadline.**

→「避免……」可以「avoid sth. from happening」表示，為固定用法。而相關的片語in case在意義上有些不同，表示「以免……」，其常用的句型有「in case of sth. 」，或是「in case +子句」。

③ **speed up our progress 加速進度**

→progress一詞表「進度；進展」，同時有動詞和名詞詞性。郵件中的make faster progress為名詞用法之一；而替代句則為另一種名詞用法speed up progress。

④ **put people's lives in danger**

→要表達「因……危及……」除了可以使用「risk sth.」以外，也可以使用「put sth. in danger」，表示「把……置於危險之中」。

♥ **hasty** *adj.* 倉促的；輕率的
♥ **preparation** *n.* 準備
♥ **avoid** *v.* 避免
♥ **in advance** 預先；事先
♥ **apply for** 申請
♥ **cut corners** 偷工減料；貪便宜

馬上來練習吧！

想要表示進度確認時該怎麼說呢？

1. 開頭稱謂　Mr. Cooper,

2. 問候句　I am writing to ______________________________
____
_____________________________________.

3. 信件主要內容　__________________________, We planned to have
____;
however, ______________________________________
____
____
___________________________________.

4. 結尾問候句　Best regards,

5. 署名　President

參考解答 Answer

Mr. Cooper,

I am writing to confirm the progress of teaching plan this semester.

According to the school schedule, we planned to have midterm on October 15th this semester; however, due to the delay on school progress that brought by the typhoon last week, we will have to postpone the exam to the week after that.

Best regards,

President

中譯

庫柏先生您好：

此來信是為了向您確認本學期的教學計畫進度。

根據學校行事曆，我們原定在十月十五日舉行期中考。不過，因為上星期的颱風造成進度的延後，我們將把考試延期至原定的下個星期。

校長 敬上

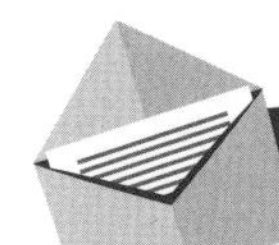

句型這樣替換也可以

1. **We planned to have…** 我們原定要……
2. **we will have to postpone the…to…** 我們將把……延期至……

Unit 01 講解訂購流程

寄信時可以這麼說

Dear Madam,

We learn from the advertisement that your latest **multifunction** printer, CY-5100, **is highly received** in our country. This model can ① not only print documents but also photos, copy, scan, and fax ②. This is just what we need. To make our first **transaction** with you **smooth** ③, we'd like to request ④ information about how we can **place an order** for 2,000 of that printer.

Your prompt reply will be greatly appreciated.

Sincerely,

Edward Hoffman

您好：

我們從廣告上得知貴公司最新型的CY5100**多功能**印表機在國內**非常暢銷**。這款機型不但能①列印文件，還能影印照片、備份、掃描，和傳真。這正是我們所需要的②。為了讓我們與您的第一次**交易**能夠**順利進行**③，我們想詢問④一下該如何**下訂**2000台該款印表機。

若您能盡快回覆，我們將不勝感激。

愛德華‧霍夫曼 謹上

換句話說

① **is able to / is capable of**

→表達「能力」除了用基本的can以外，還可以用be able to或be capable of等片語，增加內容與用語的深度。

② **Besides printing documents, this model can also print photos, copy, scan, and fax.**

→not only...but also...片語視固定用法，表示「不但……還……」，而相似用語besides也可用於此處，因此這句可以替換為besides printing documents, this model can also print photos, copy, scan, and fax.。

③ **successful**

→交易除了可以「順利進行」以外，也可以用「成功」形容，因此可替換為to make our first transaction successful.。

④ **know about**

→郵件中的最後是想讓對方知道要下訂單的想法，並詢問方法，request一詞便是表「詢問」。而此處也可以直接使用we'd like to know about...也是詢問的一種方式。

單字片語急救包

♥ **multifunction** *adj.* 多功能的
♥ **transaction** *n.* 交易
♥ **smooth** *adj.* 順利進行的
♥ **be highly received** 非常暢銷
♥ **place an order** 下訂單

回覆時可以這麼說

Dear Mr. Hoffman,

Thank you for your inquiry regarding ① our multifunction printer CY-5100.

The purchase order form is attached. ② Please print it out and fill it out as specifically as possible ③ and then send it to us by fax. We will **check out** our **inventory** to see ④ if we have enough **stock on hand** before sending you the **purchase confirmation form**.

Any interests in other products please contact Mr. Collins at our Service Department directly.

Sincerely yours,

Sonix Co. Ltd

霍夫曼先生：

感謝您諮詢①本公司的CY5100多功能印表機。

附上訂購單的檔案②，請列印出來，盡可能明確③填妥後傳真到本公司。我們會在寄**訂購確認單**給您之前先**核對庫存**，以查看④是否**手邊**有足夠的**存貨量**。

若您對其他產品有興趣，請直接聯絡本公司客服部的柯林先生。

索尼克斯有限公司 謹上

換句話說

① showing an interest in

→此類回信的方式，通常會感謝原寄件人的詢問，例如郵件中的首句thank you for your inquiry...；另外，還可以感謝對方對自己公司產品的興趣，也就是替換句thank you for showing an interest in...。

② The attachment is the purchase order form.

→郵件中若有附帶附加檔案，通常有兩種方式表達。一是使用動詞片語be attached；二是使用名詞片語the attachment be...。因此，本句可替換為the attachment is the purchase order form.。

③ in detail

→specifically一詞表示「詳盡地」，也可以用in detail代替，因此本句可替換為fill it out in detail。

④ make sure

→make sure片語表示「確認」，可於此處替換原句的see。

單字片語急救包

♥ **inventory** *n.* 庫存
♥ **stock** *n.* 存貨
♥ **purchase confirmation form** *n.* 採購確認表
♥ **check out** 查看；核對
♥ **on hand** 手邊；旁邊

馬上來練習吧！

想要講解訂購流程時該怎麼說呢？

1. 開頭稱謂　Dear U & Me Shopping Mall,

2. 問候句　I placed an order ______________________________
____
______________________________.

3. 信件主要內容　I am writing to tell you that ______________.
____
____
____
______________________________.

4. 結尾問候句　I will ______________________________
____
______________________________.

5. 署名　Faithfully,

Helen Kim

參考解答 *Answer*

Dear U & Me Shopping Mall,

I placed an order for a couple of items from your online shopping mall a few days ago, and my purchase just arrived this morning.

I am writing to tell you that I am very pleased with your service. Your commodities are fairly good and the delivery of goods is prompt.

I will definitely continue purchasing from your online shopping mall.

Faithfully,

Helen Kim

中譯

親愛的優與美購物中心：

我在幾天前在您的線上購物中心訂了幾樣商品，今天早上我買的東西剛送到。我是寫來告訴您，我對貴公司的服務感到相當滿意。你們的商品非常的好，而且運送也很快速。

我一定會繼續在你們的線上購物中心買東西的。

海倫．金 謹上

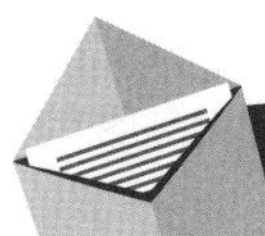

句型這樣替換也可以

1. **Thank you for your previous assistance.** 感謝您先前的協助。
2. **I just placed another purchase order and faxed the list to you.** 我剛剛又下了另一筆訂單並已傳真給您。
3. **I'm more than happy to be writing this letter to you to show my excitement about the goods.** 我非常樂意寫這封信向您告知我對收到產品的興奮之情。

Unit 02 說明商品運送事宜

寄信時可以這麼說

Dear Mr. Burns,

Thank you for purchasing our products.

I am writing to inform you that we provide ① **home-delivery** service, and it is **free of charge** ②. Your **merchandise** will be delivered ③ to the destination you **indicate** in three days ④ after the order is confirmed. Please **specify** the time you'd like your merchandise to arrive clearly on the order form.

Should there be any questions, please do not hesitate to contact us.

Best regards,

Chloe Kern

伯恩斯先生：

感謝您購買本公司的產品。

在此通知您本公司提供①**免費**②**宅配到府**服務。您的**商品**將在訂單確定後三天內④送達③您**指定**的地點。請將您希望物品送達的時間**清楚地註明**在訂貨單上。

若有任何問題，請不吝諮詢。

克蘿伊・柯恩 謹上

換句話說

① offer

→provide和offer兩個動詞都表示「提供」，所以提供服務可以用provide service和offer service，兩種皆很常用。

② complimentary

→在買賣中常見到的用語free of charge為「免費」之意，也可以用形容詞complimentary一詞表示。

③ sent / shipped

→貨物的運送除了可以用deliver一詞以外，sent和ship也可以和merchandise、cargo、commodity、goods等「貨物」搭配使用。

④ in 24 hours 二十四小時內／in a week 一週內

→「in + 時間」片語可以表示「在……時間內」，因此，此處還可以用in 24 hours（二十四小時內），或in a week（一週內）等替換。

- ♥ **home-delivery** *n.* 宅配到府
- ♥ **merchandise** *n.* 商品；貨物
- ♥ **indicate** *v.* 指定
- ♥ **specify** *v.* 具體說明；詳細指出
- ♥ **free of charge** 免費

回覆時可以這麼說

Dear Chloe,

I have just faxed you the order form on which the **ship to address** and the time we're available to receive ① the goods ② are specified.

I'd like to **make a request** that since the **commodities** I ordered ③ from you are **delicate** and **fragile**, please make sure they are well-wrapped **in case** they are damaged ④ **in transit**.

Your **considerate** service is greatly appreciated.

Warmest regards, Barry Burns

克蘿伊小姐：

我方才已經將訂購單傳真給您，上面清楚註明了**寄送地址**以及我們方便收①貨②的時間。

因為我向貴公司訂購③的**商品**是**精緻易碎**品，所以我想**請**您務必妥善包裝之，**以免**在**運送過程中**有所損壞④。

感謝您**貼心的**服務。

貝瑞．伯恩斯 謹

換句話說

① pick up

→pick up除了可以表示最常用的「接（某人）」以外，還有很多其他意義，例如：pick up the phone（接電話）、pick up some milk in the grocery store（到雜貨店買一些牛奶），或者pick up the goods（收貨）。

② merchandises / commodities

→表示「商品；貨物」的名詞有很多，這裡整理一下剛剛出現過的所有名詞：merchandise、goods、cargo、commodity。值得注意的是，其中的goods一般情況都是用複數形式。

③ purchased

→郵件中的order表示「訂購」，另一種比buy正式的「購賣」動詞為purchase，因此此處可以作替換詞。

④ in case of doing damage to them

→damage一詞表「損壞；破壞」，有名詞和動詞兩種形式。作動詞時，用法為damage sth.，郵件中的be damaged就是其被動形式；作名詞時，用法為do damage to sth.，因此此處可以替代為in case of doing damage to them。

單字片語急救包

- ♥ **ship to address** *n.* 運送地點
- ♥ **commodity** *n.* 商品
- ♥ **delicate** *adj.* 脆弱的；精緻的
- ♥ **fragile** *adj.* 易碎的
- ♥ **considerate** *adj.* 細心的
- ♥ **make a request** 請求
- ♥ **in case** 以免
- ♥ **in transit** 運送途中

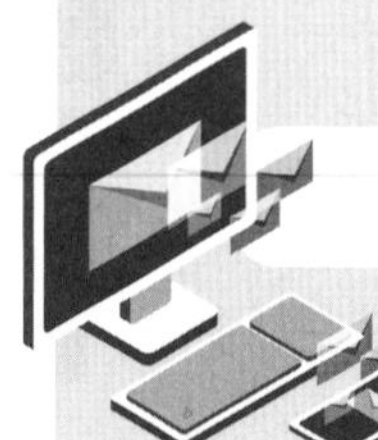

馬上來練習吧！

想要說明商品運送事宜時該怎麼說呢？

1. 開頭稱謂 Dear Service Department,

2. 問候句 I have ordered ______________________________
____
______________________.

3. 信件主要內容 I would like to know ______________________.
If it is possible, can I request that ____________
____
____
____
_________________?

4. 結尾問候句 Please ______________________________.
Thank you very much.

5. 署名 Yours,
Christina Edward

參考解答 *Answer*

Dear Service Department,

I have ordered a dozen of goblets on the Internet from your company. I would like to know how you usually ship fragile items like these. If it is possible, can I request that you ship my goblets by home-delivery service?

Please notify me at your earliest convenience. Thank you very much.

Yours,

Christina Edward

中譯

客服部您好：

我已經在網路上向貴公司訂購了一打高腳杯。我想知道您們通常是如何運送像這樣的易碎物品。如果可能的話，我是否能要求您以宅配服務運送我的高腳杯呢？

方便的話請盡早通知我。非常感謝您。

克莉絲汀娜．愛德華　謹上

句型這樣替換也可以

1. **Thank you for purchasing...** 感謝您購買……
2. **Your merchandise will be delivered to...** 您的商品將會送達至……
3. **Please specify the time you'd like...** 請您註明希望……的時間。

Unit 03 詢問商品寄送細節

寄信時可以這麼說

Dear Mr. Morris,

I intend to ① order a couple of household appliances ② from your company, but before I **place the order**, I'd like to inquire how my merchandise will be delivered ③. Also ④, I would like to know if the **delivery charge** will be **itemized** on my **invoice**. I truly love the design of your products, and I hope the ordering procedure wouldn't be a problem. Thank you for your help.

I am looking forward to your early reply.

Yours,

Cindy Holmes

墨里斯先生：

我打算①向貴公司訂購數樣家電用品②，但在我**下訂單**之前，我想詢問一下我的商品將會以何種方式運送③。此外④，我想知道**運費**是否將會被**列**在我的**收據**上。我真的很喜歡你的產品的設計，希望訂購過程不會是個問題。謝謝你的幫助。

期待您盡早回覆。

莘蒂．荷姆斯 謹上

換句話說

① plan to

→郵件中的intend to表示「打算做」，也可以替換為plan to，一樣表示做某事的打算、念頭。不過相較之下，intend to會比plan to更正式一點。

② a few pieces of furniture 幾件傢俱／ 1,000 pairs of leather boots 一千雙皮靴／ 350 KGs of charcoal 三百五十斤的木炭

→這裡的句型為「單位+物品」。值得注意的是，要特別留意單位量詞的用法。傢俱furniture可以用piece，靴子boots可以用pair，木炭charcoal可以用重量單位KG。

③ the shipping method of my merchandise

→郵件中原句欲表達的是「想知道商品的運送方式」，直接使用shipping method意詞，代表「運送方式」之意。

④ what's more

→句首的Also為轉接詞，用來暗示下一個句子的話題會有所轉換。此處還可以用what's more替換使用。

單字片語急救包

♥ **delivery charge** *n.* 運費
♥ **itemize** *v.* 分條列舉、詳細列述
♥ **invoice** *n.* 收據；發票
♥ **place an order** 下訂單；訂購

回覆時可以這麼說

Dear Ms. Holmes,

Generally ①, we have all our commodities **shipped** ② to our customers ③ by **freight** ④ unless there are some special requests. We will pay **transportation** fees and **insurance** directly to the **carrier**, and then itemize these charges on your invoice.

If there's still anything we can **be of service to** you, please let us know.

Best regards,

David Morris

親愛的荷姆斯女士：

一般來說①，本公司是以**貨運**的方式④將商品**運寄**②給顧客③，除非有其他特別要求。 我們會直接支付**運費**及**保險費**給**運輸公司**，然後將這些費用詳細列舉在您的帳單上。

如還有任何我們可以**為您效勞**之處，敬請賜知。

大衛．墨里斯 謹上

換句話說

① Normally / Usually

→郵件中的generally為發語詞，放在句首使用，表示「一般來說」，也就是泛指所有的情況。除此之外，normally和usually也有相同的詞意，可以替換使用。

② we ship all our commodities

→郵件中的用法和替換句的差異只在於主被動的形式。have sth. shipped表示「讓……被運送」，也可以直接用主動式ship sth.，簡化句型。

③ clients 客戶／purchasers 購買者／consumers 消費者

→在商業買賣上，表示「買方」的用詞有多種，除了郵件中的customer顧客以外，還有其他的用詞，例如：客戶client、購買者purchaser、消費者consumer。

④ by express delivery 以快遞方式／by ship 海運／by air 空運

→運送方式除了郵件中的貨運by freight以外，還有其他種方式，例如：快遞 by express delivery、海運by ship，或空運by air。

單字片語急救包

- ♥ **ship** *v.* 運送
- ♥ **freight** *n.* 貨運；貨物
- ♥ **transportation** *n.* 運費
- ♥ **insurance** *n.* 保險
- ♥ **carrier** *n.* 運輸公司；運送人
- ♥ **be of service to** 為某人效勞

馬上來練習吧！

回覆詢問商品寄送細節時該怎麼說呢？

1. 開頭稱謂

Dear Ms. Edward,

2. 問候句

In reply to ____________. In addition, ____________
____
____.

3. 信件主要內容

Thus, ____
____
____
____
________________.

4. 結尾問候句

Please feel free ______________________________
____
________________________.

5. 署名

With regards,

Service Department

參考解答 Answer

Dear Ms. Edward,

In reply to your letter dated February 13, we are pleased to notify you that door-to-door service is the consistent style of our company. In addition, we always specifically instruct the carriers to handle fragile items with extra care.

Thus, I can assure you that your merchandise will arrive in good condition.

Please feel free to contact us if you still have any questions.

With regards,

Service Department

中譯

親愛的愛德華小姐：

回覆您二月十三日的來信，我們很樂意通知您，宅配到府是我們公司一貫的作風。 再者，我們通常會特別指示運送員格外小心地處理易碎物品。

因此，我可以向您保證，您的商品會完好無缺的送達。

若您還有任何問題，請不吝諮詢。

客服部 謹上

句型這樣替換也可以

1. I'd like to make a request that... 我想請您……

2. I intend to order... 我打算訂購……

3. I would like to know if... 我想知道是否……

Unit 04 詢問試用意願

寄信時可以這麼說

Dear Mr. Walker,

We are pleased to announce ① the **launch** of our new product, a multifunction printer, which **integrates** a printer, a copier, a scanner and a fax **into** one **single** machine.

Any opinion ② from our potential clients is really important and helpful for us. We appreciate it if you would be interested in **trying** it **out**.

The **item sample** will be presented ③ to you immediately upon your favorable response ④.

Looking forward to hearing from you soon.

Yours sincerely,

Haggard **Corporation**

沃克先生，您好：

很高興向您宣佈①本公司**開發**了一台多功能事務機，將印表機、影印機、掃描機及傳真機**合而為一**。

任何潛在客戶的意見②對我們來說都很重要也很有幫助，若您願意③**試用**，我們將不勝感激。

一經您給予肯定的回覆，**樣品**便會立刻寄去④給您。

期待您的回覆。

海格**公司** 謹上

換句話說

① declare / release

→launch一詞在此表「發表會」，通常會與announce「宣布；發布」一詞連用，除此之外，declare與release也是常用詞，在此可以做替換。

② review

→review一詞除了表示「複習」之外，也可以表示「評論」，因此通常也會使用於顧客意見、回饋等處，例如：film / book / user review（電影／書籍／使用者評論）。

③ sent

→此處的present表示「贈送；授予」，也可以用sent（寄送）一詞替換。值得注意的是，此處的動詞應使用過去分詞（p.p），表示被動語態，因為item sample是被動「被送」給對方的受詞，而sent一詞為send的過去分詞形式。

④ as soon as we get your favorable response

→郵件中此處欲表達「只要一收到收信人的同意回覆，就會寄送樣品」，因此可以用「as soon as + 子句」，也就是as soon as we get your favorable response替代。

單字片語急救包

♥ **launch** *n./v.* 發表會；發起；推出
♥ **single** *adj.* 單一的；單身的
♥ **item sample** *n.* 樣品
♥ **corporation** *n.* 集團公司
♥ **integrates ...into...** 將……合而為一；結合
♥ **try sth. out** 試用

回覆時可以這麼說

Dear Madam,

We are honored to ① have the opportunity to try out your new product. Your products are always **satisfying** ② and never **let** users **down**. We would appreciate it if you would send us your sample. Also, please attach an **instruction manual** as well as ③ the price list as a **reference** for us to ④decide whether to purchase after using it.

We will contact you when we receive them.

Thank you very much.

Best regards,

Oliver Walker

敬啟者：

我們很榮幸①能有機會試用貴公司的新產品。貴公司的產品總是很**令人滿意**②，從不**讓**使用者**失望**。若您能寄送樣品過來，我們將不勝感激。同時，請附上**使用說明書**和③價目表，如此我們才能在使用後**參考**④是否要購買。

我們收到東西時，會再通知您的。

非常感謝您。

奧立佛 · 沃克 謹上

換句話說

① are pleased to

→在商業的書信往來，表示「我們很榮幸……」或是「……是我們的榮幸」是很常見的語句。為了向對方表示自己的尊敬，通常除了用be honored to外，常見的用法還有be pleased to。

② user-friendly 易於使用的／environment-friendly 友善環境的

→郵件中的satisfying一詞表示「令人滿意的」，可以用來形容商品特徵的詞彙還有user-friendly（易於使用的），和environment-friendly（友善環境的）等。

③ along with

→as well as一詞表示「也；和」，連接對等的兩個詞或片語。同樣地，along with也是連接詞，可以在此處做替換。

④ so that we can

→「so that + 子句」能表示目的，因此郵件此處可以改為so that we can decide whether to...，表示「如此，我們就能決定是否要……」。

- ♥ **satisfying** *adj.* 令人滿意的；令人滿足的
- ♥ **instruction manual** *n.* 使用說明書
- ♥ **reference** *n.* 參考
- ♥ **let sb. down** 使……失望

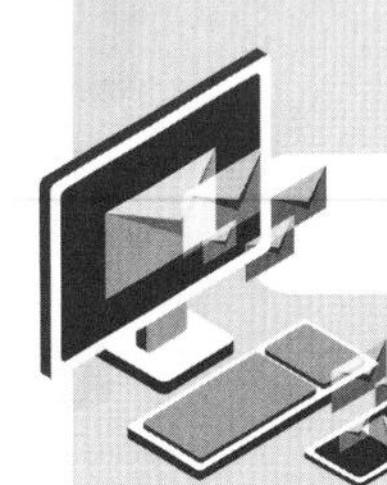

馬上來練習吧！

想要詢問適用意願時該怎麼説呢？

1. 開頭稱謂

Dear Sir,

2. 問候句

We learned from the advertisement that ________
____.
As ______________________________________
________________________________.

3. 信件主要內容

Could you kindly __________________________
____
____
____
__________________________?

4. 結尾問候句

Hoping to receive your early reply.

5. 署名

With thanks and regards,

Judy Franklin

參考解答 Answer

Dear Sir,

We learned from the advertisement that you produce various fabrics of excellent quality. As a manufacturer who makes garments of all kinds, we are in need of large quantities of various fabrics.

Could you kindly send some fabric samples as well as the price list and color swatches to us?

Hoping to receive your early reply.

With thanks and regards,

Judy Franklin

中譯

您好：

我們從廣告上得知貴公司生產各式各樣高品質的布料。身為製造不同種類服裝的廠商，我們需要大量不同的布料。

能否請您惠賜一些布料樣品，以及價格表和色樣給我們呢？

希望能盡快得到您的回覆。

感謝並祝福您。

茱蒂・法蘭克林

句型這樣替換也可以

1. **We appreciate it if you ...** 若您……我們將不勝感激
2. **This will not only... but also...** 這不但會……而且……
3. **It would be our honor to have this opportunity to try out your new product.** 若能試用貴公司的新產品將會是我們的榮幸。

Unit 05 請求寄送樣品

寄信時可以這麼說

Dear Sir,

I have learned from your advertisement that your company produces ① electronic **components of superior quality** ②, so I would like to request ③ some **typical** samples of your products. We are **holding a meeting** next week to select the items we will ④ purchase in the next year, so hopefully you could send us the samples by this Friday.

I am looking forward to hearing from you soon.

Sincerely,

TSUC Electric Co.

敬啟者，您好：

我從廣告上得知貴公司的廣告中得知您生產①**高品質的**電子**零件**②，因此想請③您提供貴公司產品中一些具**有代表性的**樣品。我們下週將**開會**選出下一年度所要④訂購的產品，因此希望您能夠在本週五之前將樣品寄來給我們。

希望能很快得到您的回覆。

TSUC電子公司 謹上

換句話說

① **manufacture 生產；製造**

→製造一詞除了郵件中的produce以外，另一個常見的詞為manufacture。另外，-er表示「做某事的人／者」，所以製造商可以用producer或manufacturer表示。

② **superior-quality / high-end / cutting-edge electronic components**

→「of + Adj + N」可以表示某事物的特徵，因此郵件此處的...of superior quality表示「高品質的……」。同樣的特徵，也可以直接將形容詞放置於名詞前修飾，例如：superior-quality products。其他還有和高科技相關的形容詞，例如：high-end（高檔的；高價位的），和cutting-edge（領先的；尖端的）等。

③ **ask for**

→ask for wth.片語表示「（向某人）請求某物」，可以在此處替換request sth.。另外，ask的用法還有「ask to + V」，表示「（向某人）請求做某事」，例如：Linda asked me to pick her up at 6.（琳達請我在六點時接她）。

④ **are going to**

→表示未來將要做的事，最常使用的兩個方式就是will和be going to，所以此處也可以做互相替換。不過，值得注意的是，這兩種方式並不是完全一樣的，不是在所有的情況下都能做替換。簡而言之，will通常偏向「意見；猜測」，例如：Polar bears will be endangered in the near future.（北極熊即將面臨絕種）；而be going to則是比較基於「現有的事實」推測的，因此可信度相對比較高，例如：It is cloudy. Maybe it is going to rain.（烏雲好厚，等等應該就會下雨了）。

單字片語急救包

♥ **component** *n.* 零件；組成部分
♥ **typical** *adj.* 經典的；有代表性的
♥ **of superior quality** 品質優異的
♥ **hold a meeting** 開會

回覆時可以這麼說

Dear Madam,

We are very pleased to send you our samples.

I am writing to advise ① you that we have shipped the samples you requested by express delivery ② today. It will arrive in about ③ 3 days. I have also attached our **catalog** as well as the price list and the color **swatches**. You will find out that our product **is the most** outstanding one among all the competitors' ④ after using it.

Please inform us when you receive them. Should you need anything else, please also let us know.

I will be looking forward to your **feedback**.

Sincerely,

Tom Green

敬啟者：

我們非常樂意提供樣品給您參考。

特此通知①您，您所要求的樣品已經在今天下午以快遞②寄送過去了。大約③三天內就能送達。我同時附上了本公司產品的**型錄**、價目表及顏色**樣本**。試用過後，你會發現我們的商品**是最**出色**的**④。

您收到樣品時，麻煩通知我們一下。如果還需要其他東西，也請讓我們知道。

期待您給我們看過樣品的**心得**。

湯姆．葛林 謹上

換句話說

① inform

→郵件中此處的advise譯為「通知」而非「建議；勸告」，因此還可以用常見的inform做替換。而inform的用法又分兩種，一為inform sb. of sth.，二為inform sb. that + 子句。

② by freight 貨運／by post 郵寄

→郵件中的by express delivery為「以快遞（的方式）」，除此之外，還有其他種運送方式，例如：貨運by freight或郵寄by post等。

③ around

→表示「大約」之意，about和around兩個詞都可以，不過比較起來，前者的頻率會比後者高一點。另外，若是用於估計數字以外的範圍，則是後者較常使用，例如：all around the world（全世界）和beat around the bush（說話拐彎抹角）等。

④ stand out from the rest

→表示「表現傑出；表現突出」除了可以用郵件中的句子外，也可以轉換out-standing的詞性為動詞片語stand out from sth.，也就是stand out from the rest (of the competitors' products)。

單字片語急救包

♥ **catalog** *n.* 目錄；型錄（= catalogue）
♥ **swatch** *n.* 樣品
♥ **feedback** *n.* 回饋
♥ **be the most...** 是最……的

馬上來練習吧！

回覆請求寄送樣品時該怎麼說呢？

1. 開頭稱謂　Dear Ms. Franklin,

2. 問候句

Thanks for ______________________________.

We have pleasure ________________________.

____.

_________________________.

3. 信件主要內容

Please ___________________________________

____

____

____

________________________________.

4. 結尾問候句

We are looking forward ______________________

_______________________________________.

5. 署名

Best regards,

C & Q Co.

參考解答 Answer

Dear Ms. Franklin,

Thanks for being interested in our products. We have pleasure in sending you our sample fabrics. The samples were shipped to you today by express delivery. I have also attached a catalogue of our products as well as a price list.

Please inform us when you receive them.

We are looking forward to your feedback.

Best regards,

C & Q Co.

中譯

法蘭克林女士：

感謝您對本公司的產品感興趣。我們很高興能寄送我們的布料樣品給您。樣品已經在今日以快遞寄送給您了。我同時附上了本公司的商品目錄以及價格表。

收到東西時請通知我們。

我們十分期待您對樣品的感受。

希爾克公司 謹上

句型這樣替換也可以

1. **We are honored to have the opportunity to...**
   我們很榮幸能有機會……
2. **We will contact you when...** ……時我們會通知您。
3. **I am looking forward to...** 期待著您……

Unit 01 | 抱怨商品延遲到貨

寄信時可以這麼說

Dear Mr. Miller,

I **regret to** tell you that I haven't received the items I ordered from your company, which should have arrived two weeks ago ①. I inquired your Service Department by telephone ② yesterday but didn't get a **definite** reply. I would like you to confirm the **status** of my order #203456831 dated April 15.

Please make prompt **response** and solve ③ the problem immediately. This is a very disappointing ④ experience.

Sincerely,

Olivia Cage

米勒先生：

我**很遺憾地**要告訴你，我向你們公司訂購，兩星期之前①就應該送到的商品到現在還沒收到。我昨天打電話②到貴公司的客服部詢問，卻沒有得到**明確的**回答。我想請你確認我在四月十五日訂購的編號203456831 貨品目前的**狀況**。

請迅速給我**回覆**，並立即處理③這個問題。這是個令人失望的④經驗。

奧利薇・凱吉

換句話說

① three days ago 三天前／last month 上個月／yesterday 昨天／already 已經

→郵件中的two weeks ago表示「兩星期前」，也就是說「時間 + ago」表示「……以前」。除此之外，還可以用「last + 時間單位」表示過去的時間點「上個……」，例如：last month。或者是用yesterday（昨天）和already（已經）表達。

② by e-mail 用電子郵件／by fax 用傳真

→郵件中的by telephone表示「用電話（的方式做某事）」，所以替換方式可以依據所需的聯絡途徑，在by後加入即可。

③ deal with

→常與problem一詞連用的動詞除了有solve（解決）一詞外，還可以用deal with片語，表示「處理；解決」。另外，problem solved.常被當作獨立的句子使用於口語中，表示「問題解決了」，而deal with則無這種用法。

④ unsatisfying 不令人滿足的／unpleasant 不開心的

→表示某件事物沒有達到自己的期望可以用disappointing（令人失望的），或是unsatisfying（不令人滿足的）和unpleasant（不開心的）。

♥ **definite** *adj.* 肯定的；明確的
♥ **status** *n.* 狀況；情形
♥ **response** *n.* 回應
♥ **regret to** 遺憾地做某事

回覆時可以這麼說

Dear Ms. Cage,

We were **embarrassed** to discover that there's been a **mix-up** over the dates ①. Please accept my sincere ② apology for our carelessness and the inconvenience that may have caused you.

We would like to inform you that your order has been sent and you will receive it by 5:00 p.m. today. **To make up for** ③ our mistake, we also sent a free gift for you, hoping for your forgiveness ④.

We have a **taken measures** to **ensure** that such mistake does not happen again.

Truly yours,

Kyle Miller

親愛的凱吉小姐：

我們**很慚愧**地發現我們將日期①**搞錯**了。因為我們的粗心造成您的不便，在此致上萬分的②歉意。

我們要通知您，您訂購的商品已經寄出，您會在今天下午五點之前收到貨。**為了補償**③此次的疏失，我們還一同寄送了免費禮物，希望能得到您的原諒④。

我們已經**採取相關措施**，**保證**此等情事不會再發生。

凱爾・米勒 謹上

換句話說

① ship to addresses 送貨地址／deadlines 交貨限期

→此處提及的是「出錯的問題」，除了郵件中的date（日期）之外，其他像是ship to addresses（送貨地址）和deadlines（交貨日期）也可以多留意一下。

② genuine / earnest

→表示道歉的誠意，常用的形容詞有sincere（有誠意的），或是genuine（真誠的）和earnest（誠摯的）。除此之外，這類形容詞有時也可以用於感謝他人時，例如：sincere / genuine / earnest appreciation。

③ to compensate for

→表示「彌補；補償」之意，可以用make up for sth.片語，或是to compensate for sth.片語，兩種可以互相替換。另外，若要用名詞表示，則可以使用compensation一詞。

④ to restore our reputation 以恢復我們的聲譽

→hope一詞表「希望；祈求」，其主要用法有兩種。一為hope for sth.，如郵件中的hoping for your forgiveness（希望能得到您的原諒）；二為hope to V.表示「希望能（做到……）」，因此此句可改為，hope to restore our reputation（希望能恢復我們的聲譽）。

- ♥ **embarrassed** *adj.* 慚愧的；困窘的
- ♥ **mix-up** *n.* 搞混
- ♥ **ensure** *v.* 確保；保證
- ♥ **to make up for sth.** 補償；彌補
- ♥ **take measures** 採取措施

馬上來練習吧！

想要抱怨商品延遲到貨時該怎麼說呢？

1. 開頭稱謂　Dear Amanda,

2. 問候句　______________ we ordered from you ______________; however, ______________.

3. 信件主要內容　Please ______________ ______________ ______________ ______________ ______________.

4. 結尾問候句　We would like to ______________ ______________ ______________.

5. 署名　Sincerely,

Debbie Stevenson

參考解答 Answer

Dear Amanda,

The LED light bulbs we ordered from you on August 19 (Purchase Order #3035829) were supposed to be delivered three days ago; however, we haven't received them up to now. Please track the goods at once.

We would like to hear from you regarding this problem immediately.

Sincerely,

Debbie Stevenson

中譯

親愛的艾曼達：

我們在八月十九日向您訂購的LED電燈泡（訂單編號3035829）三天前就應該送到了，然而我們到目前為止都還沒收到貨。請立刻追蹤貨物。

我們希望能立刻得到您對此問題的回覆。

黛比．史蒂文生 謹上

句型這樣替換也可以

1. **I would like you to confirm the status of...**
   我想請您確認……的狀況
2. **Please make prompt response** 請迅速回覆
3. **I am awfully sorry for the late delivery of...**
   對於……延遲，我感到非常抱歉

Unit 02 針對商品延遲到貨致歉

寄信時可以這麼說

Dear Ms. Bloom,

We are greatly ① sorry to inform you that **due to** ② a mix-up at our freight company ③, there will be a late delivery of your products ④. Please be **assured** that we will work with a **reliable** freight company to make sure this doesn't happen again ⑤.

Again, I apologize for any inconvenience this delay may have caused.

Sincerely yours,

KTC Co.

親愛的布魯姆小姐：

很①抱歉在此通知您，**由於**②運輸公司③的失誤，您的貨品將會延遲送達④。我們**保證**將會與**可靠的**運輸公司合作，以避免這樣的事件再度發生⑤。

再一次為此次延誤將造成的不便向您致歉。

KTC公司 謹上

換句話說

① **extremely 極度地／deeply 深深地／terribly 非常地**

→郵件中這裡的greatly是修飾sorry的副詞，表示感到抱歉的程度。此外，還可以用extremely（極度地）、deeply（深深地），或是terribly（非常地）連用。值得注意的是，表示感謝和表示歉意的副詞不一定能通用。舉例而言，像terribly一詞通常只與負面的動詞或形容詞搭配使用。

② **owing to / because of**

→表示事發的原因最常使用的片語除了due to以外，owing to和because of都是一樣的用法。值得注意的是because一詞的用法為二，其一為because of sth.，二為because that + 子句。郵件中的用法則屬第一種。

③ **over the delivery deadlines 在交貨限期上／over the quantities of your order 您訂單的數量上**

→郵件的此處欲表示「發生錯誤的地方」，也就是運輸公司。另外，「交貨期限」和「訂單數量」的寫法分別為over the delivery deadlines和over the quantities of your order。

④ **your products will be postponed for a few days 您的貨品將會延遲幾天送達**

→郵件中的there will be...是書信中常用的表達方式，也可以直接用一般句型「S（主詞）+ V（動詞）+ O.C（受詞補語）」表示，替換成your products will be postponed for a few days。

⑤ **avoid this from happening again**

→郵件中的make sure...表示「確認（某事不會發生）」，另外可用「避免（某事發生）」的反面方式表達，也就是avoid...from happening，因此替代句為avoid this from happening again。

♥ **reliable** *adj.* 可靠的；可信的
♥ **assure** *v.* 向……保證；讓……放心
♥ **due to** 由於

回覆時可以這麼說

Dear Sir,

As you may know ①, time of delivery is **a matter of great importance** to us. We **are in urgent need of** the goods because they are **demanded** by our customers ②. I feel **reluctant** to tell you that we can't wait any longer.

If you can't meet the delivery date, I am afraid that we may have to cancel the order ③. Thank you for your understanding.

Also, thanks to your acknowledgment in advance so that we can make adjustment ④ early.

Faithfully yours,

Amber Bloom

先生您好：

您也知道①，交貨時間對我們來說是**非常重要的一件事**。我們**急需**這批貨，因為客戶們已經在**催了**②。**很抱歉**我們無法再等了。

如果貴公司無法在交貨日前交貨，恐怕我們就得取消訂單了③。敬請諒解。

另外，感謝您的及早發現，讓我們也能盡早做出變動④。

安珀・布魯姆 謹上

換句話說

① **As a matter of fact 事實上；說真的**

→郵件中的as you may know片語可視為發語詞，通常用於句子的一開始，表示「如您所知；您也知道」以開啟話題。而這裡也可以用as a matter of fact片語，表示「事實上；說實話」之意。

② **we have an intensive schedule 我們的行程安排緊湊**

→郵件此處欲表達「因為商品是他們客戶要的，所以原訂的日期不能延遲」，換句話說，他們時間的安排是固定不能變動的，也就是很緊湊之意，可以用we have an intensive schedule替換表示。

③ **turn down your reschedule 拒絕您的改期**

→turn down片語表示「拒絕某事」，因此此處可以用turn down your reschedule替換。另外，turn down表示拒絕的用法還有：turn down an invitation（拒絕邀約）或turn someone down（拒絕某人）等。

④ **alteration 變動／another arrangement 另外安排**

→表示「（計畫的）改變」除了可以用adjustment（更動；變動）之外，也可以用alteration（變動）或是another arrangement（另外安排）表示。

單字片語急救包

- ♥ **demand** *v.* 要求；需求
- ♥ **reluctant** *adj.* 不情願的；勉強的
- ♥ **a matter of great importance** 非常重要的事
- ♥ **be in urgent need of** 急需要某物

馬上來練習吧！

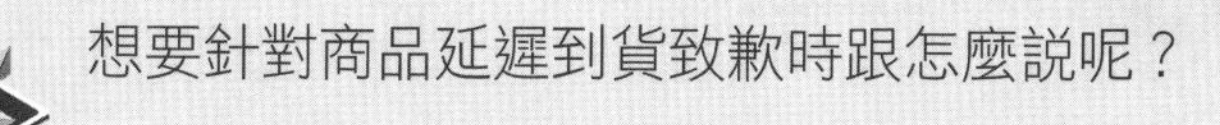

想要針對商品延遲到貨致歉時跟怎麼說呢？

1. 開頭稱謂

Dear Debbie,

2. 問候句

In reply to your letter ______________________,
____
______________________________.

3. 信件主要內容

I am sure ______________________________________
____
____.
____
____________________________.

4. 結尾問候句

We apologize for _______________________________
____.

5. 署名

Best regards,

Amanda Douglas

參考解答 Answer

Dear Debbie,

In reply to your letter dated September 10, I have checked our delivery schedule, and found that your order was shipped on September 11. I am sure there was a mix-up over the dates at the freight company. The good news is that you should receive the goods within today.

We apologize for any trouble that the delay may have caused.

Best regards,

Amanda Douglas

中譯

親愛的黛比：

回覆您九月十日的來信，我已經查過我們的出貨時程表，並發現您訂的貨在九月 十一日已送出。我確定運送公司那邊把日期搞錯了。好消息是您應該在今天之內就收到貨品了。

我們為延遲到貨可能造成的任何困擾向您致歉。

艾曼達・道格拉斯 謹上

句型這樣替換也可以

1. **We were embarrassed to discover that...** 我們很慚愧地發現……
2. **Please accept my sincere apology for...** 對於……請接受我們誠摯的歉意
3. **We have taken measures to ensure that...** 我們已經採取相關措施以保證……

Unit 03 協助退貨

寄信時可以這麼說

Dear Customers,

We regret to announce that our dishwasher ①, model no. DW-3580, has been **found to be defective** ②. If you have purchased this product, please call us at 707-3588. Our **Customer Service** will assist ③ you with the return shipping details.

Please accept our **genuine** apologies for the inconvenience this may have caused.

Thank you for supporting us ④.

Sincerely yours,

JQB Appliance Co.

親愛的顧客您好：

很抱歉通知您，本公司型號DW-3580的洗碗機①出現了一些**問題**②。如果您買了這項產品，請致電 707-3588，本公司的**客服人員**將會協助③您退貨事宜。

造成您的不便，在此向您致上**最深的**歉意。

謝謝您對本公司的支持④。

JQB 電器公司 謹上

換句話說

① **hair dryer 吹風機／vacuum cleaner 吸塵器／heater 暖氣機／air-conditioner 冷氣機／cooker 電爐／hanger 衣架**

→有些器具或機器的形式為「-er」，例如郵件中的dishwaher，其他還有像是：吹風機hair dryer、吸塵器vacuum cleaner、暖氣機heater、冷氣機air-conditioner、電爐cooker和衣架hanger等，皆為此種方式。

② **malfunctioning 機能失常的／out of order 故障的**

→郵件中的defective一詞表示「有缺陷的」，除了可以用來形容機器或商品以外，也可以用來表示人的個性不完美。除此之外，malfunctioning和out of order也都有一樣的意義，可以在此做替換。

③ **help**

→assist一詞表示「幫忙；協助」，後面會與介系詞with連用，而help也有一樣的用法，help / sb. with sth.為此處使用的片語，兩字可互相做替換。

④ **purchasing our products 謝謝您購買我們的商品／visiting for website 謝謝您光臨我們的網站**

→在買賣交易的書信往來中，賣方常會在信的最後謝謝買方的購買或任何其他付出，例如：purchasing our products（謝謝您購買我們的商品），或是visiting for website（謝謝您光臨我們的網站）。

單字片語急救包

- ♥ **defective** *adj.* 有缺陷的；有問題的
- ♥ **Customer Service** *n.* 客服
- ♥ **genuine** *adj.* 由衷的
- ♥ **found to be…** 被發現是……

回覆時可以這麼說

Dear Madam,

In reply to your letter dated November 10, I want to tell you that I am very pleased with ① the service we have received. Your honesty is really **invaluable** in the market nowadays ②, **not to mention** your kind attitude toward customers ③. Thank you very much for your prompt **resolution** of this defective product problem. We will surely continue to purchase goods from you ④.

With thanks and regards,

Vincent Chang

小姐您好：

回應您十一月十日的來信，我想告訴您，我對我們得到的服務感到非常滿意①。您的誠信在當今②的市場上是非常**可貴的**，**更不用說**是您們對顧客良好的態度了③。感謝您如此迅速的**解決**此一瑕疵產品問題。我們一定會繼續向您購買商品④。

感謝並祝福您。

張文森 謹上

換句話說

① **to express my gratitude for 表達我的感謝**

→郵件中，寄件人欲傳達他對原寄件人的感謝之意。除了原句的用法外，也可以用替代句I want to express my gratitude for...。

② **in these days 當今／in this generation 在這個世代**

→表示「當今；現在」除了可以用nowadays以外，也可以用in these days或是in this generation（在這個世代）。

③ **the way you dealt with the problem 你處理問題的方式**

→郵件中的attitude toward sth.是常用片語，表示「對於……的態度」，這邊想說明的是賣方對於顧客的態度很好。除此之外，此處也可以用替代句the way you dealt with the problem表示他們處理問題的方式（很恰當）。

④ **support your company 支持貴公司**

→郵件最後提到，買方很認同賣方的作法，所以會繼續購買他們的產品。換句話說，買方會繼續對賣方的公司給予支持，所以可以用替代句continue to support your company。

♥ **invaluable** *adj.* 可貴的
♥ **resolution** *n.* 解決
♥ **in reply to** 作為答覆
♥ **not to mention** 更不用說

馬上來練習吧！

想要協助退貨時該怎麼說呢？

1.	開頭稱謂	Dear Anna Pole,
2.	問候句	In reply to your letter of October 29, ___________ ____ ___________________________.
3.	信件主要內容	We will surely _____________________ ____ ____ ____ _________________________.
4.	結尾問候句	We will try our best to rectify this situation, and will get back to you soon.
5.	署名	Yours sincerely, Service Department

參考解答 *Answer*

Dear Anna Pole,

In reply to your letter of October 29, we are very sorry to hear that your customers found this product unsatisfactory; however, this upright vacuum cleaner never receives the same complaints from any of our other dealers. We will surely test its suction performance and launch an investigation on the problem.

We will try our best to rectify this situation, and will get back to you soon.

Yours sincerely,

Service Department

中譯

親愛的安娜・坡：

回覆您十月二十九日的來信，我們很遺憾聽到您的客戶發現這個產品令人不甚滿意的消息；然而，這款直立式吸塵器從未接獲其他經銷商相同的控訴。我們一定會測試其吸力性能，並且展開針對此問題的調查研究。

我們將盡全力處理這個情形，並會馬上與您聯繫。

客服部 謹上

句型這樣替換也可以

1. **We have received...** 我們已經收到了……
2. **Can I return and exchange it for...** 我可以將它拿回去換成……嗎
3. **I was terribly sorry to know that...** 得知……我感到非常抱歉
4. **We will certainly accept the return of...** 我們當然接受……的退貨
5. **We regret to announce that...** 很抱歉通知您……
6. **If you…please call us.** 如果您……請打電話給我們。

Unit 04 投訴商品毀損

寄信時可以這麼說

Dear Carrie,

I am writing to inform you that our order arrived timely ① this morning, but unfortunately, we found quite a few ceramic vases ② chipped on the **rims**. I believe the chips were caused ③ **in transit** owing to the **random** packing.

We simply ④ can't accept the **damaged** items, and will return them to you **COD** for a refund.

Sincerely,

Sam Jefferson

親愛的凱莉：

在此通知您我們訂的貨物今天上午已經及時①送達了，但是很遺憾地，我們發現有不少陶瓶②**邊緣**有缺口。我想這些缺口應該是包裝**不完整**，在**運送過程中**產生的③。

我們實在④無法接受這些**受損的商品**，因此將以**貨到付款**的方式退還給您，並要求退費。

山姆．傑佛森 謹上

換句話說

① on time / well-timed

→郵件中的timely表示「準時的」，同樣可以使用on time或well-timed兩個片語。值得注意的是，on time和in time常常容易被搞混。兩個片語的意義分別為「準時地」和「及時地」，後者有「及時趕上（剛好壓線）」之意，例如：he braked his car in time, so no one got hurt.（他及時剎住他的車，所以沒有人受傷）。

② crystal plates 水晶盤／porcelain bowls 瓷碗／glass cups 玻璃杯

→郵件中的ceramic vase為「陶瓶」，其他還有容器用品的名稱如：水晶燈crystal plate、瓷器porcelain bowl，和玻璃杯glass cup等。

③ the damage was done / caused

→通常「毀損；破壞」常會用damage一詞表示，而常與damage搭配使用的動詞有do和cause，所以此處可以替代為the damage was done / caused in transit。

④ really / totally

→郵件中的simply表示「實在；絕對」之意，而非「輕易地；簡單地」，同樣意義地單字還可以用really或totally修飾，以加強「無法接受」的程度、態度和語氣。

- ♥ **rim** *n.* 邊緣
- ♥ **random** *adj.* 隨機的
- ♥ **damaged** *adj.* 受損的
- ♥ **in transit** 運輸途中
- ♥ **COD (=cash on delivery)** 貨到付款

回覆時可以這麼說

Dear Mr. Jefferson,

We're deeply sorry to hear that the ceramic vases you ordered were partly damaged during the **transportation** ①. This situation seldom ② happens in our company.

As it is our **responsibility** to deliver ③ goods **in good condition**, we certainly will refund the money. However, we must remind you that these products are only **refundable** in a specific ④ period.

Sorry again for all the inconvenience.

Sincerely,

Carrie Jones

親愛的傑佛森先生：

聽到您訂購的陶瓶**在運送過程中**①有部分受損，我們深表遺憾。這種狀況在我們公司是很少②發生的。

將貨物以**良好狀態**送達③是我們的**責任**，因此我們當然會退費給您。然而，在此必須提醒您，這些商品只**能**在特定④期間內**退貨**。

再次為所有的不便致歉。

凱莉·瓊斯 謹上

換句話說

① in the duration of transportation

→during一詞表示「在……的期間」，用法如郵件中during+時間範圍即可。另一用法是轉化詞性為名詞duration，用in the duration of +時間範圍表示，因此此句可以替換為during the transportation。

② rarely / scarcely

→表示「很少；幾乎不」時，除了可以用seldom以外，rarely和scarcely也是常用的頻率副詞，用來修飾做某動作的次數或頻率，因此在此可以互相替換使用。

③ provide / offer

→郵件中，此處表示「將貨品以良好了狀態送達是我們的責任」，而此處的「送達」也可以直接用「提供」表示，因此provide和offer兩個動詞皆可以在此處做替換。

④ regulated

→郵件中，此處欲表達的意思為「特定的時間內（才可退貨）」，也就是說，時間範圍是有「明確規定的」，因此也可以用regulated一詞，表示「規定的」之意。

單字片語急救包

♥ **transportation** *n.* 運送
♥ **responsibility** *n.* 責任
♥ **refundable** *adj.* 可退還的
♥ **in good condition** 狀態良好的

馬上來練習吧！

想要投訴商品毀損該怎麼說呢？

1. 開頭稱謂　Dear Mr. Sullivan,

2. 問候句　________________ has arrived, but I regret to inform you that ______________________________
____
______________________________.

3. 信件主要內容　I would like to know ______________________
____
____
____
__________________.

4. 結尾問候句　I appreciate ____________________________
__________________________________.

5. 署名　Regards,
Josephine

參考解答 Answer

Dear Mr. Sullivan,

The ornamental glass we ordered from your company has arrived, but I regret to inform you that several pieces of them were broken during the transportation. I would like to know whether we need to send them back COD for replacements.

I appreciate your immediate reply.

Regards,

Josephine

中譯

蘇利文先生：

我們向貴公司訂購的裝飾玻璃已經送到了，但是很遺憾我必須通知您，其中有幾件 在運送途中損毀了。我想知道我們是否需要將它們以貨到付款的方式寄回，以更換新品。

您若能盡速回覆，將不勝感激。

喬瑟芬 謹上

句型這樣替換也可以

1. **We simply can't understand...** 我們實在無法理解……
2. **If there are any questions, please...** 如果有任何問題，請……
3. **We are reluctant to announce that...** 雖然很不願意，但我們仍必須告訴您……
4. **we are now negotiating with...** 我們目前正在跟……協商
5. **We fully understand that...** 我們完全理解……

Unit 05 協調更換新品

寄信時可以這麼說

Dear Mr. White,

We are reluctant to announce that the file cabinets ① which you ordered have been found damaged ②. To provide you with the commodities **of the best quality**, we are now **negotiating with** our **supplier** for new **replacement**.

Consequently ③, we're sorry to inform you that your merchandise will be delayed in delivery. Please allow 10-14 ④ days to receive your order. We appreciate it very much for your understanding.

Sincerely,

Angela Black

親愛的懷特先生：

雖然很不願意，但我們仍必須告訴您，我們發現您所訂購的檔案櫃①是缺損②品。為了提供您**品質最好的商品**，我們目前正在**跟供應廠商協調更換**新品。

因此③，很抱歉在此通知您，您的商品將會暫時延遲送貨。請您等待十至十四天④。非常感謝您的諒解。

安琪拉・布萊克 謹上

換句話說

① **the wardrobes 衣櫃／the table lamps 檯燈／the bedside cupboards 床頭櫃**

→郵件中的file cabinet為「檔案櫃」，其他相關的傢俱用品如：衣櫃wardrobe、檯燈table lamp、床頭櫃bedside cupboard寫法可以注意一下。

② **flawed 有瑕疵的**

→通常商品貨物有缺陷、瑕疵時，除了可以用damage一詞表示，也可以用flaw一詞替換，flawed products可以表示「瑕疵品」。若要表達「完美無瑕」，則可用形容詞flawless表示。

③ **So / Therefore**

→郵件中的consequently表示「因此」，其名詞形式為consequence表示「結果；後果」。另外，so和therefore兩個詞也可以在此作替換，用法和原句一樣至於句首，後加逗點隔開主要子句。

④ **additional shipping**

→additional一詞表示「額外的；附加的」，因此此處可以替換為please allow additional shipping days to receive your order.（請您等待額外的運送時間）。值得注意的是，商業說法常常出現的「工作天」，會以business day表示。

單字片語急救包

♥ **supplier** *n.* 供應商
♥ **replacement** *n.* 替代；更換
♥ **consequently** *adv.* 因此；所以
♥ **of the best quality** 品質最好的
♥ **negotiating with** 與……協商

回覆時可以這麼說

Dear Ms. Black,

We fully ① understand that this delay in delivery is not your fault ②, and I am writing to tell you that we are able to give you ③ an **extension** of two weeks. Please **make certain** this does not happen again. However, we would like to request a 10% **discount** on this merchandise because of the delay in delivery ④.

Please send me a reply at your earliest convenience.

Yours sincerely,

Timothy White

布萊克小姐您好：

我們完全①理解這次的延遲並非貴公司的錯②，因此在此告知您，我們同意給予③您兩星期的**延期**交貨。請**確保**這種事情不會再發生。然而，因為送貨延遲④，我們想要求你們給予這批貨物九折的**優惠**。

請您盡快給我回覆。

堤莫西・懷特 謹上

換句話說

① totally

→understand一詞常和fully連用，修飾「理解的程度」很「完全」，另外還可以一起搭配使用的副詞還有totally，一樣表示「完全的；全面的」之意，因此可以在此作替換使用。

② the quality of commodities is often unpredictable 商品的品質很難預測

→郵件中，此句欲表達「出貨延遲是無法避免的」，換言之，也就是說是因為「商品的品質很難預測」the quality of commodities is often unpredictable，所以才會造成這樣的延遲。

③ allow 允許

→allow一詞表示「允許」，隱含著命令或是規定的口吻，通常會用在上對下的對話中。因為郵件中的延遲是賣方的問題，所以買方在這也可以用allow表示，「允許給予對方兩星期的延期時間」。

④ as a compensation 作為補償

→商業往來的溝通中，常常會談及賠償、補償的內容，這時可以用動詞compensate表示「彌補」之意，或是用名詞形式...as a compensation（作為補償）表示。因此此處的替代句表示「要求九折的優惠以作為貨物延遲的補償」。

♥ **extension** *n.* 延期
♥ **discount** *n.* 優惠；折扣
♥ **make certain** 確定

馬上來練習吧！

回覆協調更換新品時該怎麼說呢？

1. 開頭稱謂　Dear Josephine,

2. 問候句　First of all, ______________________________
____
______________________________.

3. 信件主要內容　We will ____________________________
____.
____,
____
______________________.

4. 結尾問候句　Again, we ____________________________
____.

5. 署名　Regards,
Hank Sullivan

參考解答 Answer

Dear Josephine,

First of all, I must apologize for our carelessness during the shipment. We will certainly exchange your goods. We will have our man send you the new replacements and retrieve the damaged ones within the next two days, so it is unnecessary that you send them back to us COD.

Again, we apologize for causing you all the trouble.

Regards,

Hank Sullivan

中譯

親愛的喬瑟芬：

首先，我必須為我們運送過程中的疏忽致歉。我們一定會為您更換貨品。我們會在這兩天內派人將新品送去，並取回毀損的物件，所以您沒有必要將它們以貨到付款的方式寄回給我們。

再次為造成您的麻煩致歉。

漢克．蘇利文 謹上

句型這樣替換也可以

1. **We are reluctant to announce that...** 雖然很不願意，但我們仍必須告訴您……
2. **we are now negotiating with...** 我們目前正在跟……協商
3. **We fully understand that...** 我們完全理解……

Part 3

英文E-mail 實例大全
— 部門篇

Chapter

Unit 01 | 詢問是否增加人力配置

寄信時可以這麼說

Mark,

As you know, it is time for our annual university **recruiting drive** ①. Due to ② recent **budget cuts**, we will only be **taking on** seven new **interns** as a result of this drive. I know that you are running ③ **short-staffed** at present ④, so I need to know if you have any **staffing requirements** for the next year.

Thanks.

Gladys Cloethra

馬克：

如你所知，現在是我們每年的大學**徵才活動**時間①。由於②最近的**預算削減**，所以我們將只能**延攬**七名新進**實習員工**。我知道你們部門目前④是在**人員編製短缺**的情況下運作③，所以我需要知道你們明年度是不是有**人員配置需要**。

謝謝。

葛萊迪．可羅沙

換句話說

① **it is time that we recruit for university / it is time for us to recruit for university**

→與it is time相關的片語使用有很多種，其一為it is time for + N.，也就是郵件中原句的使用方式；另外，替代句則分別為it is time that + 子句和it is time for sb. to + V.的使用。

② **Given / Owing to**

→郵件中的due to表示「由於；出於某原因」，同樣可以使用owing to或是given方式替換。值得注意的是，以分詞當句首時，若是動作為被動，需用過去分詞。這裡的budget cuts指的是遭到給予的動作，應為被動，所以不能用現在分詞giving。

③ **operating 營運**

→郵件中的run表示「運作」，而非「跑步」，因此在此可以用operate一詞替換，同樣可表示公司的運作、營運。

④ **presently**

→郵件中的at present表示「目前；當下」，也可以轉換詞性，改用副詞presently表示。值得注意的是，這裡的時間副詞除了放在句尾，也可以往前移到be動詞後，也就是改為you are presently running short-staffed。

單字片語急救包

- ♥ **recruiting drive** *n.* 徵才活動
- ♥ **budget cut** *n.* 預算刪減
- ♥ **intern** *n.* 實習生
- ♥ **short-staffed** *n.* 人員編製短缺
- ♥ **staffing requirement** *n.* 人員編製需求
- ♥ **taking on** 承擔

回覆時可以這麼說

Gladys,

Thank you for thinking of us. Presently we can **manage with** our existing ① staff **complement**. Our employees are always hard-working ② just to keep our department running smoothly. However, if there is an intern available ③ after the other departments have had their requirements met, we will have an upcoming project where we could definitely use an extra hand ④.

Regards,

Mark Morrison

葛萊迪：

感謝你還想到我們。目前，我們可以用現有的①人員編制**設法應付**過去。我們的員工都很努力工作②，讓我們部門運作順暢。但是，如果其他部門的人員編製需求都已經達到，而還有多一個實習生名額③的話，我們即將有個案子要開始做，肯定會需要多個人手幫忙④。

馬克．莫里森 謹上

換句話說

① **current**

→郵件中的existing表示「現有的；現存的」，使用範圍很廣，例如：Under the existing condition the global warming is only getting worse.（在目前的情況下，全球暖化只會惡化）。另外，current也可以在此替代使用，通常會用來表示「現任的」，例如：current president（現任總統）。

② **working industriously**

→郵件中的hard-working一詞為形容詞，表示「工作勤奮的」。此外，這句也可以用working + Adv.表示，例如替代句的working industriously。由於industrious一詞表示「勤奮的」，所以副詞形式加-ly可以用來修飾work的狀態。

③ **vacancy 空缺；缺額**

→提及「缺額」時，英文中通常會使用vacancy一詞表示。例如郵件中的實習生名額可以改為intern vacancy，以及職場上常出現的「工作職缺」也會用job vacancy表示。

④ **would be in need of extra hand**

→片語in need of sth.可以用來表示「需要某物」，因此此處可以更改為we would be in need of extra help，表示「需要更多人手幫忙」。

♥ **complement** *n.* 整數；整體

♥ **manage with** 用……設法應付

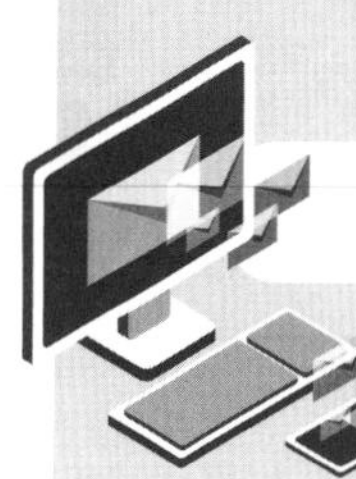

馬上來練習吧！

想要詢問是否增加人力配置時該怎麼說呢？

1. 開頭稱謂　To the Shipping Manager–New York,

2. 問候句　We ______________________________
____
__________________________.

3. 信件主要內容　Could you let me know ______________
____
____
____
__________________.

4. 結尾問候句　Thanks.

5. 署名　Laura Able

參考解答 Answer

To the Shipping Manager–New York,

We have been made aware by management that the shipping department across all our offices is understaffed. Could you let me know what your requirements for additional staff will be in the next while? I am preparing to publish nationwide advertising to recruit for this shortage.

Thanks.

Laura Able

中譯

致紐約運務經理：

我們已從管理部門那裡知道我們所有辦事處的運務部皆是處在人員配備不足的狀態。能否請您讓我知道在未來一段時間內您需要增加的員工人數？我目前正在針對這個人員短缺問題，準備發佈全國性的招募廣告。

謝謝您。

羅拉．亞伯 謹上

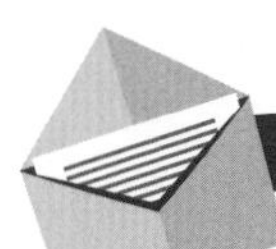

句型這樣替換也可以

1. **It is time for us to...** 我們該……了
2. **Due to...we will...** 由於……，我們將……
3. **I need to know if...** 我需要知道是否……

Unit 02 詢問空缺補滿時間

寄信時可以這麼說

Dear Audrey,

You may **be aware that** we recently signed ① a new contract with Acme Consulting for the development of an online billing **interface** ②. Given the deadlines on this project and other **ongoing** work, I will be needing ③ an additional two ④ **developers** to join my team. I know this is a **short notice**, but can we try and get these positions filled before the end of the month?

Regards,

John Falkland

親愛的奧黛莉：

您可能已經**知道**我們最近跟 Acme顧問公司簽①了一份開發線上收費**裝置**②的合約吧。由於要趕這個案子的完成期限和其他**正在進行的**工作，我會另外④需要③兩名④**開發人員**加入團隊。我知道這**通知**很**倉促**，但我們是否能在年底前將這些職位補滿呢？

約翰·佛克蘭

換句話說

① enter into

→「簽約」除了可以用sign a contract以外，替代句的enter into a contract也是英文中會使用的用法。反之，若要表達「違反合約」，則可以用break a contract片語表示。

② the production of a large amount of computer components 大宗電腦零件的生產／the construction of a large-scale shopping center 一間大型購物中心的建設

→郵件中的合約是developing of an online billing interface「開發線上收費裝置」，其他可能用到的詞彙像是生產production ，或是建設construction的相關用法可以參考替代句。

③ requiring

→郵件中的need表示「需要」，同樣意義的詞可以用require代替，因此此句可以替換為I will be requiring...。

④ another two / two other

→郵件中的an additional two developers表示「額外的兩個開發人員」，也可以用「額外」改用「另外」表示，也就是替代句的another two或是two other。值得注意的是，兩個看起來很相似的詞another和other的用法不能隨意交換。

單字片語急救包

- ♥ **interface** *n.* 介面；聯繫
- ♥ **ongoing** *adj.* 進行中的
- ♥ **developer** *n.* 開發人員
- ♥ **short notice** *n.* 匆促的通知
- ♥ **be aware that/of** 注意到；意識到；知道

回覆時可以這麼說

Dear John,

Thanks for the request. I will **start the ball rolling** on this ① immediately; however, I do need a couple ② things from you.

Can you have Edwards Moore please **sign off** on this request as all new staff ③ need to **go through** him.

Also, please send through a detailed **job description** so I can get it posted ④ **ASAP**.

Thanks.

Audrey

約翰：

謝謝您提出要求。我將**開始著手進行**這件事情①；但是，我有兩件②事情要您幫忙。

麻煩您請愛德華．摩爾**批准**這項需求，因為所有新進工作人員③，都必須**經**他**同意**。

另外，請給我一份關於**工作內容的詳細描述**，好讓我**盡快**公佈④這項人事需求。

謝謝。

奧黛莉

換句話說

① get on with this

→郵件中的start the ball rolling on sth.表示「開始進行某事」，同樣的意義除了用常見的動詞start表達以外，還有get on with sth.片語可以使用。

② a few

→郵件中的a couple表示「一對；幾個」，通常是指兩個左右的數目。這裡也可以用a few代替，同樣表示「一些；幾個」之意。值得注意的是，a few和few是不同的意義。few一詞為否定用法，通常用來說明「幾乎沒有」的數量。兩個詞的用法分別如下：he has a few friends.（他有一些朋友）；he has few friends.（他很少朋友）。

③ employees

→郵件中的staff表示「員工；工作人員」，還可以用雇員employee表示。值得注意的是，staff一詞為「全體員工」的統稱，所以通常用單數使用；而employee在此則要加上複數s。

④ announced

→郵件中的動詞post表示「公布；發布」，通常暗指有實際上「廣告、資訊張貼」的動作；而announce則沒有強調具體的動作，可能只有口頭上的公布。

單字片語急救包

- ♥ **job description** *n.* 職務描述
- ♥ **start the ball rolling** 開始進行
- ♥ **sign off** （不需簽字的）批准、同意
- ♥ **go through** 被通過
- ♥ **ASAP** 盡快、馬上（=as soon as possible）

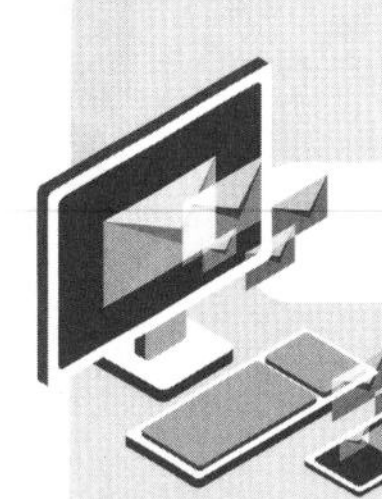

馬上來練習吧！

想要詢問空缺補滿時間時該怎麼說呢？

1. 開頭稱謂　Dear Laura,

2. 問候句　Thank you so much for the email.

3. 信件主要內容

____.
At my location, we ____________________
____.
____
____
______________.

4. 結尾問候句　Thanks.

5. 署名　Graham Earnest

參考解答 Answer

Dear Laura,

Thank you so much for the email. I am very happy to see that our recent difficulties have been recognized and addressed. At my location, we have an immediate need for two shippers who have experience in international freight as well as an additional general laborer. Forklift driving experience will be a definite bonus for all positions.

Thanks.

Graham Earnest

中譯

親愛的羅拉：

非常感謝您的來信。知道我們最近遇到的困難已經被注意到並積極處理，我感到非常高興。在我的部門，我們急需兩位有國際運務經驗的託運員，還有一位非專業性的工人。所有職務中，有駕駛起貨機的經驗會更好。

謝謝您。

格拉漢・爾尼斯特

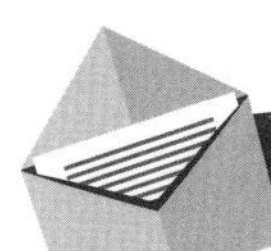

句型這樣替換也可以

1. **Thank you for thinking of...** 感謝你想到……
2. **You may be aware that...** 您可能已經知道……
3. **Thanks for the request of** 謝謝您提出……的要求

Unit 03 考勤異常

寄信時可以這麼說

To all staff members,

Recently we have seen increased levels of **absenteeism** ①. I would like to **draw everyone's attention** to ② the following, taken from our **employee handbook**.

- All incidents of **sick leave** require either a **doctor's letter**, or **prior approval** from your manager.
- Any incident without prior approval from a manager or no letter from a doctor **is treated as** absenteeism and no pay will be given for that day.

All the employees are asked to follow ③ the above rules. Nobody should be an exception. ④

Sincerely,

Gwyneth Smith

致全體員工：

最近我們發現員工**曠職**的情形提高了①。我想請大家**注意**②以下事項，此乃出於我們的**員工手冊**。

- 所有**病假**都需要**醫生證明**，或**事先**得到你的經理**批准**。
- 任何未經經理事先批准或沒有醫生證明的情況將**被視為**曠職，當天不給付薪資。

所有的員工皆要遵守③以上規定，沒有人例外。④

官妮絲・史密斯 謹上

換句話說

① **there has been an increase in the level of absenteeism**

→郵件中的increased為形容詞，表示「上升的；增加的」，而若要使用increase的名詞形式，可以替換為have an increase in sth.片語，因此本句可以改為there has been an increase in the level of absenteeism。

② **everyone to pay attention to**

→attention一詞表示「專注；注意力」，相關用法除了郵件中原句的draw one's attention以外，也可以用常用的pay attention to sth.片語表示，例如：Please pay attention to the class; stop chatting.（請專注於課堂上，不要聊天）。

③ **comply with 遵守**

→表示「遵守」除了使用常見的follow一詞以外，comply with也是書信中常用的片語。

④ **Not any special privilege is granted. 沒有人能享有任何特權**

→郵件中，此句欲表達「沒有人是例外（不用遵守規定）」，換言之，也就是「沒有人能享有特權」，因此可以使用替代句Not any special privilege is granted.表示。

單字片語急救包

- ♥ **absenteeism** *n.* 曠職
- ♥ **employee handbook** *n.* 員工手冊
- ♥ **sick leave** *n.* 病假
- ♥ **doctor's letter** *n.* 醫生證明
- ♥ **prior approval** *n.* 事先核准
- ♥ **draw one's attention** 吸引某人注意
- ♥ **be treated as** 視同

回覆時可以這麼說

Dear Gwyneth Smith,

Thank you for **clearing up** ① the company **policy**. I **happen to** have a related matter so I'm writing to ask for your suggestion.

I have been recently **involved in** a car accident and am unable to attend work every day due to numerous ② **physio therapy** appointments and follow up x-rays. I have been working from home during this time. Can you please let me know how I deal with ③ this within ④ the company's policy?

Thanks in advance.

Lonnie Lauren

親愛的官妮絲．史密斯：

謝謝你對公司**政策**的**清楚說明**①。我**剛好**有相關的問題想詢問你的意見。

我最近**發生了**一場車禍，但因為要做多次②**物理治療**以及照追蹤X光而無法每天去上班。我這段時間一直在家工作。能否請你告訴我該如何以④公司政策處理③這個問題？

在此先謝謝你了。

羅里．羅倫

換句話說

① your detailed explanation on 您詳盡的解釋

→郵件中的clear up表示「釐清；弄清楚」，在此也可以用「詳細的解釋」表達，因此此句可以替換為Thank you for your detailed explanation on the company policy.。

② a series of / a battery of 一連串的

→在醫療相關的主題，常常會提及「一連串的治療」字眼，通常會用a series of examination或是a battery of examination表示。此處一樣屬醫療相關的主題「物理治療」，因此同樣可以使用a series of和a battery of 修飾physio therapy。

③ handle

→表示「處理」除了可以用郵件中的deal with片語以外，也可以使用handle一字表示，因此此句可改為...how I handle this within the company's policy。

④ without violating

→郵件中，原句欲表達「如何公司政策的範疇內處理」，換言之，就是「以不要違背公司規定的前提下處理」，因此可以替換為deal with this without violating the company's policy。violate一詞表示「違背」。

單字片語急救包

- ♥ **policy** *n.* 政策
- ♥ **physio therapy** *n.* 物理治療
- ♥ **clearing up** 弄清楚；澄清
- ♥ **happen to** 恰巧……
- ♥ **involve in** 涉及；牽涉於

馬上來練習吧！

想要表示考勤異常時該怎麼說呢？

1. 開頭稱謂　To All Staff,

2. 問候句　Following ____.

3. 信件主要內容　I would like to ____.

4. 結尾問候句　Regards,

5. 署名　Beverley Orange

參考解答 *Answer*

To All Staff,

Following the recent introduction of our work from home policy, we have seen some uneven staffing levels in a number of our departments. I would like to reinforce with all staff that the flextime policy we have does still mean everyone needs to be in the office from 10am to 3pm everyday. An additional 3 hours of the work day can be completed at home; however you need to co-ordinate with your manager to ensure that email and phones are answered throughout the entire business day.

Regards,

Beverley Orange

中譯

致所有同仁：

在公司最近實施在家工作政策之後，我們在一些部門中發現了人員配置不平均的情形。我想向所有員工強調，彈性工時的政策仍然要求每個人必須每天從上午十點至下午三點在辦公室上班。多出來的3個工時可以在家中完成；然而，你仍必須與部門的經理配合，保證整個工作天內所有的的電子郵件和電話皆有所回覆。

貝佛利．奧朗吉 謹上

句型這樣替換也可以

1. **I would like to draw everyone's attention to...**
   我想請大家注意……
2. **Thank you for clearing up...** 謝謝您澄清……
3. **Can you please let me know how...**
   能否請您告訴我該如何……

Unit 04 詢問績效考核

寄信時可以這麼說

To **Human Resources**,

I have been with ① the company for exactly one year ② now. During my interview and at signing on, I was informed that I would need to **undergo** an annual **performance review**.

As a **dedicated** worker, I would like to know what areas I can **improve on**. Also, I am planning an overseas vacation in the fall and would like to know approximately ③ how many days I would have available to me at that time ④.

Thanks.

Jenny Greene

致**人資部**：

我已經在①公司服務整整一年②了。在我面試和簽合約時，我被告知必須**接受**年度**績效考核**。

身為一名**專業**員工，我想知道哪些地方是我可以**改進**的。此外，我正在計畫今天秋天到海外度假，我想知道屆時④我大約③可以有多少天假期。

謝謝您。

珍妮・葛林恩

換句話說

① devoted myself to 貢獻自己於

→devote一詞表示「貢獻」，常用的相關片語為devote sth. to sth. / sb.表示「奉獻……於……」，因此此句也可以替換為devoted myself to the company表示「貢獻自己於公司」。

② a whole year

→郵件中的exactly修飾one year，表示數據相當準確，也就是「整整的；不多不少的」，而whole一詞也有相同的用法，用法很廣泛，例如：this whole thing、the whole year，或a whole day等。

③ about

→與exactly相反，要表示「大約；大概」的數量計算，可以用about或是郵件中的approximately兩個副詞修飾。使用上，通常會將about和approximately至於數量詞之前，例如：the construction will last about / approximately a year.（建設工程大約會持續一年）。

④ then

→郵件中的at that time表示「那時」，可以直接用then一詞替換。值得注意的是，在使用上，通常這兩種用法都會至於句尾。

單字片語急救包

♥ **Human Resources** *n.* 人資部
♥ **undergo** *v.* 接受
♥ **performance review** *n.* 績效考核
♥ **dedicated** *adj.* 盡心盡力的；盡責的
♥ **improve on** 改進

回覆時可以這麼說

Dear Jenny,

I have **taken** this **up** ① **with** your manager, and he will be **conducting** your annual review with you within the next week ②. As regards to ③ you vacation, you will have 17 days available to you ④. Please **clear** all vacation requests **with** your manager before **submitting** them to me for **processing**.

Regards,

William Vorbryt

珍妮：

我已經**向**你的經理**提出**①這個問題，他會在下週內②**執行**你的績效考核。至於③你的假期，你會有十七天可利用④。請在**得到**經理**批准**所有的休假申請之後，再**提交**給我**辦理**。

威廉．佛布里特

換句話說

① **brought up 提出；提起**

→表示「提出」除了如郵件中的take sth. up以外，也可以用bring sth. up片語表示，例如：Don't bring this up now.（現在別提到這件事）。另外，若把片語改成bring sb. up則表示「養育；撫養某人」。

② **within the next two weeks 在未來兩週內／around the end of the month 在月底左右／by July 31 在七月三十一日以前**

→郵件中的within the next week 表示「下週內」，其他時間表達方式，例如：within the next two weeks（在未來兩週內）、around the end of the month（在月底左右），或by July 31（在七月三十一日以前）也可以在此替換使用。

③ **In regard to / with regard to / regarding / concerning**

→表示「關於；有關」的詞彙及片語有很多，除了郵件中使用的as regards to以外，還有In regard to、with regard to、regarding和concerning可以表達。

④ **to utilize / to take advantage of**

→郵件中的available通常表示「可用的；有空的」，若要直接表達「可利用的」則可用to utilize或to take advantage of兩個片語。

♥ **conduct** *v.* 處理；實施
♥ **submit** *v.* 遞交；提交
♥ **process** *v.* 辦理
♥ **take up sth. with sb.** 向某人提出某事
♥ **clear sth. with sb.** 獲得某人批准

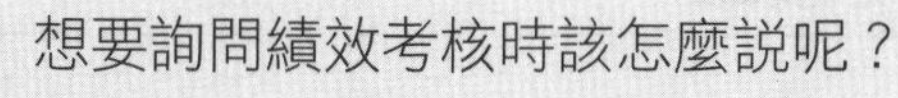

馬上來練習吧！

想要詢問績效考核時該怎麼說呢？

1. 開頭稱謂　Dear Beverley,

2. 問候句　I ______, ______, ______, ______, ______, ______. ______.

3. 信件主要內容　Can you confirm that ______ ______ ______ ______ ______?

4. 結尾問候句　Thanks.

5. 署名　George Holmes

參考解答 *Answer*

Dear Beverley,

I have been working from home a lot recently, and have been sending my planned work-from-home schedule through to Mike, my manager, every day. I have not got any feedback from him. However, I know that Bryan, my colleague, also works from home in the afternoon and evening. I've asked Mike about this and he says it is fine. Can you confirm that Bryan and I can continue to do this without violating company policy?

Thanks.

George Holmes

中譯

奧朗吉：

我最近很常在家工作，並每天將我的在家工作計劃寄給我的經理邁克。我從未得到他任何回應。然而，我知道我的同事布萊恩，也是下午和晚上在家工作。我問過邁克，他說沒關係。你能否確認一下，我與布萊恩繼續這麼做是否會違背公司政策呢？

謝謝你。

喬治・荷姆斯

句型這樣替換也可以

1. **I was informed that...** 我被告知……
2. **I have heard no mention of...** 我沒有聽說……
3. **I am planning...** 我正計畫……

Unit 05 詢問節慶、假日

寄信時可以這麼說

To HR,

I was wondering if we will be getting the 13th of May **off** this year. I am not sure ① if ② it is a holiday in our state; however, it is in the state where over 90% of our customers **are located**. Considering that there will be ③ nobody to help, it seems a little pointless ④ coming to work.

Cheers

Sarah Scott

致人資部：

不知道今年五月十三日那一天我們是否有**放假**？ 我不確定①那天在我們這一州是不是②假日，但是我們超過百分之九十的客戶**所在**的那一州是假日。考慮到③那天將沒有人可以協助我們，來上班似乎有點沒意義④。

祝 愉快

莎拉・史考特

換句話說

① **certain**

→表示「確定」除了可以用sure表示以外，也可以用certain，因此這裡可以替代為I am not certain about ...。另外，可以與certain一詞連用的的動詞有：feel certain、make certain、seem certain等。

② **whether**

→表示「是否」主要有兩個詞，一為郵件中使用的if，二為whether。在大部分的情況下，這兩個詞是可以互相通用的，不過下列五種情況則只適用whether：1. 以「是否」為開頭的句子。2. 至於介系詞之後。3. 至於不定詞to之前。4. 「主詞補語」的句子用whether，不用if。5. 做「同位語」而等於其前名詞之子句中，用whether，不用if。

③ **at the thought of having 一想到……**

→表示「想到某事」時，除了可以用郵件中的considering作為開頭以外，也可以用at the thought of...片語。值得注意的是，此片語後通常只接名詞或動名詞，所以此句可以改寫為At the thought of having nobody to help...。

④ **unnecessary 沒必要的／redundant 多餘的／uncalled for 不必要的**

→郵件中的最後一句欲表示「當天若還要上班很不必要」，除了向郵件中使用的seem a little pointless以外，也可以用替代詞語unnecessary（不必要的）、redundant（多餘的）或uncalled for（不必要的）等。

♥ **off** *n.* 不工作、休息

♥ **be located** 位於；所在

回覆時可以這麼說

Sarah,

To be clear ①, the 13th is not a holiday in this state, and never has been. Thus ② **it will be work as usual**. If you feel that there is **insufficient** ③ work on that day to **warrant** coming in, I suggest you discuss this with your manager. Although ④ you **have a point**, I think this discussion would still be **in vain**.

Thanks.

Wallace Regatherway

莎拉：

在此聲明①，十三日在本州並非假日，也從來不是。因此②那天**照常上班**。如果你覺得那天**沒有足夠的**③工作而沒**有**需要上班**的理由**，我建議你找你的經理討論這個問題。雖然④你說的**有道理**，但是我認為這個討論也只會是**白費力氣**而已。

謝謝。

華勒斯．里加德威

換句話說

① Just to clarify 為了澄清

→郵件中的to be clear有「為了弄清楚」之意，表示「在此聲明」，除此之外，也可以用just to clarify表示「為了澄清」，暗示後面將會提到與原寄信人想法相反的答覆。

② As a result, / Therefore,

→郵件中的thus表示「因此；所以」，同樣的意義可以用as a result或therefore表示。不過，通常這兩個片語置於句首時，後面都會用逗號隔開。

③ no sufficient / not enough / deficient in

→郵件中的insufficient一詞表示「不足的；不充分的」，由字根sufficient變化而來。因此，直接用no sufficient也可以表達一樣的意思。另外，此句還可以替換為there is not enough / deficient in work on that day。deficient一詞同樣表示為「缺乏的」，其後要接介系詞in加名詞。

④ despite (the fact) that

→郵件中的although表示「儘管；雖然」，同樣的意義可以用despite that或despite the fact that片語表示。值得注意的是，despite一詞可以單獨表示「儘管；雖然」，但其後只能加名詞，若要接子句一定要用despite (the fact) that。

單字片語急救包

♥ **insufficient** *adj.* 不充分的；不足的
♥ **warrant** *v.* 使有……的理由
♥ **to be clear** 先說好；先聲明
♥ **it will be work as usual** 將照常上班
♥ **have a point** 有道理
♥ **in vain** 白費力氣；做白工

馬上來練習吧！

想要表示詢問節慶、假日時該怎麼說呢？

1. 開頭稱謂　Dear Jane,

2. 問候句　Thanks for ______________________________
____
______________________________.

3. 信件主要內容
____.
____
____
____
____________________.

4. 結尾問候句　I would appreciate it if ______________________
______________________________.

5. 署名　John Roberts

參考解答 Answer

Dear Jane,

Thanks for updating the staff on the day off tomorrow. I just want to confirm your home and cell numbers that we have on record. I will be watching the weather closely and will call you tomorrow around lunch with a decision as to whether we will open for business the next day. I would appreciate it if you could relay that to all staff.

John Roberts

中譯

親愛的珍：

感謝您告知同仁明天停班的消息。我想確認您的住家和手機號碼。我明天會密切注意天氣狀況，並在午餐左右的時間打電話給您，以決定隔天是否要上班。屆時再麻煩您將消息發佈給所有同仁，感激不盡。

約翰・羅柏茲

句型這樣替換也可以

1. **Considering that...** 考慮到……
2. **If you feel that... I suggest you...** 如果你覺得……我建議你……
3. **Unless you hear, otherwise we will...** 除非你有聽到，否則我們……

Unit 01 新產品上市通知

寄信時可以這麼說

To our valued dealers,

We are proud to announce the **introduction** of **a** brand new **range of** cookware, **aimed at** the amateur chef ①. This range complements ② our existing **offerings** and results in ③ **a** complete ④ **suite of** kitchenware from our company. As part of the introduction of this new range, we will be sending samples to all existing **dealers**, as well as a **promotional pedestal** for displaying this new top of the line offering.

Yours Sincerely,

Ki-Chen Marketing

致我們重要的經銷業者：

我們很自豪的向各位**介紹一系列鎖定**業餘廚師①**為目標**顧客的全新炊具。這一系列搭配②我們現有的**商品**，就組成了③我們公司完整的④**一套**廚具。這次介紹這**一**新**系列**的活動部分之一，我們將會贈送樣品給所有現有的**經銷商**，還會附贈一座**促銷用的支柱**，以展示此款新系列商品。

旗艦市場部 謹上

換句話說

① **housewives 家庭主婦／small families 小家庭／singles 單身貴族**

→業餘廚師amateur chef為郵件中的目標客群，其他可能也會是潛在客群的有：家庭主婦housewives、小家庭small families和單身貴族singles。

② **completes**

→郵件中的complement為動詞，表示「補足；使完整」，同樣的意義可以使用complete一詞表示。值得注意的是，complete一詞同時有動詞和形容詞兩個詞性，此處使用的應為動詞。

③ **brings / creates**

→郵件中的result in片語表示「造成」，除了使用在此處，更常用於表示「事情的結果」，例如：The cigarette resulted in the big fire.（菸蒂造成了這場大火）。另外，相關的片語result from則表示「事情的原因」，例如：The big fire was resulted from the cigarette.（這場大火是源於一支菸蒂）。而此處也可以將results in替換為brings或creates，分別表示「帶來」和「組成；創造」。

④ **thorough**

→表示「完整的」除了可以用郵件中的complete以外，也可以以thorough表示。

單字片語急救包

♥ **introduction** *n.* 介紹；推出
♥ **offering** *n.* 出售物；商品
♥ **dealer** *n.* 業者
♥ **promotional** *adj.* 促銷的
♥ **pedestal** *n.* 基座；支柱
♥ **a range of** 一系列
♥ **aim at** 以……為目標
♥ **a suite of** 一系列；一套

回覆時可以這麼說

Dear Marketing Department,

We are thrilled with the new offerings and from our **brief** look on your website ①, this looks exactly like what we need to complete our inventory ②. However, we have a couple of questions, which I am sure you can **help** us **with**. The first and most obvious ③ is the pricing of these items as you are not **displaying** yet in the **dealer portal**. Secondly ④, is their **bulk purchase** discounting.

Regards,

Home Chef Tools Inc.

市場部您好：

我們非常喜歡這些新產品，我們**簡單的**看了一下貴公司的網站①，這一系列看起來正是我們所需要以使架上貨品完整的商品②。不過我們有幾個問題，我相信您能**給予**我們**幫助**。第一個，同時也是最明顯的③問題，即是這些項目的定價，這部分您目前還沒有**列**在**經銷商入口網站**上。其次④，則是這些商品**大宗購買**的優惠折扣問題。

家廚工具企業 謹上

換句話說

① **in your latest catalogue 你們最新的型錄／**
at your advertising brochure 你們的廣告冊子

→通常廣告資訊除了能在公司網站上找到以外，還有其他的來源，例如：型錄catalogue和廣告小冊子advertising brochure，因此本句可以做替換。

② **have been searching for so long 我們尋找已久的**

→search for sth.片語表示「尋找某物」，因此替換句this looks exactly like what we have been searching for so long則表示「這看起來就是我們尋找已久的（商品）」。

③ **noticeable**

→郵件中的obvious一詞表示「明顯的」，常常以It is obvious that + 子句連用，例如：It is obvious that you didn't sleep well last night.（你昨晚沒睡好是很明顯的事）。另外，noticeable也可以在此作替換詞。

④ **Furthermore / Additionally**

→為了方便閱讀，通常在條列式地依序提出資訊時，會將序數詞的副詞形式置於句首，不過除了firstly、secondly、thirdly這種用法外，也可以用其他方式變化，例如替換句中的furthermore和additionally也可以用於第二點或以後的敘述句中。

單字片語急救包

- ♥ **brief** *adj.* 短暫的；簡略的
- ♥ **display** *v.* 列出；展示
- ♥ **dealer portal** *n.* 經銷商的入口網站
- ♥ **bulk purchase** *n.* 大宗採購
- ♥ **help sb. with sth.** 幫助某人某事

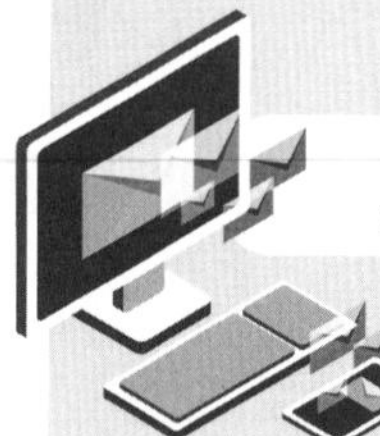

馬上來練習吧！

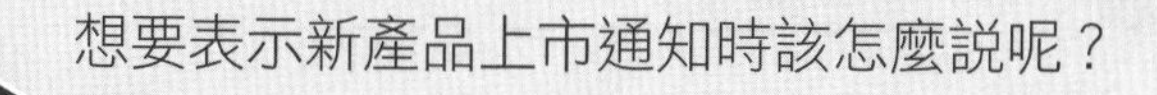

1. 開頭稱謂　Dear Partners,

2. 問候句　We are glad to ______________, ____________.

3. 信件主要內容

____,

____,

____.

____,

__________________________________.

4. 結尾問候句

Please contact us ____________________________

____

____________________________.

5. 署名

Best Regards,

Jonas Stuarts

參考解答 Answer

Dear Partners,

We are glad to introduce our latest model of the train set, TS 500, to all of you.

This new model will be on the market at the beginning of April. As most of us have been looking forward to this new product, we will also have a promotional price for this new train set. Any order placed before May 20 will receive 20% off dealer price.

Please contact us if you need more information.

Best Regards,

Jonas Stuarts

中譯

親愛的合夥人：

我們很高興向所有人介紹我們最新款的列車組TS500。

這款新車型將在四月初推出上市。由於我們大多數人一直在期待著這個新產品，我們將會提供這組新列車一個優惠價格。所有在五月二十日下達的訂單將獲得八折的交易價。

若您需要更多資訊，請與我們聯繫。

喬那斯・史都華 謹上

句型這樣替換也可以

1. **We are proud to announce that...** 我們很自豪地宣佈……
2. **we will be sending samples to...** 我們將會贈送樣品給……
3. **We are thrilled with...** 我們非常高興……

Unit 02 詢問產品上市事宜

寄信時可以這麼說

Dear Fred,

We have had numerous requests from ① our **customer base** recently about when the existing smart phone you have will be **upgraded** to one with greater memory ②. The existing device is certainly **a good seller**, and we have had a lot of positive feedback ③; however, as **alternative models** have come out ④ with more memory, we have seen our **competitors gain on** us.

Thanks.

Stephanie Beckham

佛列德：

最近我們的**客戶群**有好多人在問①什麼時候你們現有的智慧型手機，會**升級**為有更大記憶體②的手機。現有的這款當然**銷得很好**，而且我們也得到很多正面的評價③，但是因為已經有其他有更多記憶體的**替代款式**出現④，我們已經快被**競爭者超越**了。

謝謝。

史蒂芬妮・貝克漢

換句話說

① **Our products have had been widely discussed by**
我們的產品被熱烈討論
→替代語句的been widely discussd可以用於很廣泛的地方，例如：Environmental issue is a topic which been widely-discussed these years.（環境問題是近年來備受討論的議題）。

② **larger screens 更大的螢幕／thinner body 更薄的機體／faster speed 更快的速度**
→除了「更大的記憶體」以外，一般比較手機的條件還包括：更大的螢幕a larger screen、更薄的機體a thinner body，及更快的速度faster speed等。

③ **won awards 得獎**
→郵件中此句欲表示「該公司的原款手機得到很好的評價」，這裡也可以替代為won a lot of awards，表示「得了很多獎項」。

④ **appear / produced / manufactured**
→郵件中的come out片語表示「出現」，同樣的意義也可以使用appear表示。除此之外，表示「製造」的動詞produce和manufacture也可以用於此處。

單字片語急救包

- ♥ **customer base** *n.* 客戶群
- ♥ **upgrade** *v.* 升級
- ♥ **a good seller** *n.* 銷路很好的商品
- ♥ **alternative model** *n.* 替代款式
- ♥ **competitor** *n.* 競爭者
- ♥ **gain on** 逼近；超越

回覆時可以這麼說

Stephanie,

Good to hear from you, and we appreciate the positive feedback which always **encourages** us **to** improve ① our company and products. Actually ②, you are not the first to draw our attention to ③ the need to increase ④ **on board memory** on this model. As soon as the final release is out, I'll make sure to inform your company, as well as getting a **promotion display** in the mail.

Thanks.

Fred Basson

史蒂芬妮：

很高興收到您的信，也很感謝您帶給我們正面的反應，並總是**鼓勵**我們改進公司和產品①。其實②，在您之前，已經有人提醒③我們增加④這款手機**主機板記憶體**的必要性。等最後的版本一出來，我一定會立刻通知你們，同時也會在信件內提及**促銷展**的事宜。

謝謝。

佛列德・貝森

換句話說

① gives us the motivation for improving

給我們改善……的動力

→郵件中的encourage一詞表示「鼓勵；促進」，通常用在正面影響力。而此處也可以用替代句gives us the motivation for improving...表示「給我們改善……的動力」。

② In fact

→郵件中的actually表示「實際上；事實上」，除此之外，in fact片語同樣也可以用於此。不過，兩者在意義上還是有些微的差異。actually通常用於不完全同意對方的說法，或對自己前面講的那句話做些修正的時候。而in fact則是用在要對之前說的話作補充，或是舉出例證強調之時。

③ to remind us about

→remind sb. about sth.片語表示「提醒某人某事」，因此郵件中此句也可以替換為you are not the first to remind us about the need...（你不是第一個提醒我們……的必要性）。

④ enhance / enlarge / upgrade

→表示「增加；擴增某物」除了可以用郵件中的increase以外，還有替換詞enhance、enlarge和upgrade可以使用。

單字片語急救包

- ♥ **on board memory** *n.* 主機板記憶體
- ♥ **promotion display** *n.* 促銷展
- ♥ **encourage sb. to** 鼓勵某人（去做……）

馬上來練習吧！

想要詢問產品上市事宜時該怎麼說呢？

1.	開頭稱謂	Dear Jonas,
2.	問候句	Thanks ______________________________, ______________________________ __________.
3.	信件主要內容	We would like to know if ______________ ______________________________ ______________________________ ______________________________ ________________.
4.	結尾問候句	Looking forward to hearing from you.
5.	署名	Regards, Tom Ford

參考解答 Answer

Dear Jonas,

Thanks for the information of TS 500. It looks like this will be a big hit during summer season. We would like to know if there is any advertising material that you can provide for promoting this new train set, and when we will be able to place orders for TS500.

Looking forward to hearing from you.

Regards,

Tom Ford

中譯

喬那斯：

感謝您提供TS500的資訊。看起來這將會是夏季的熱銷商品。我們想知道您們有沒有任何宣傳資料可以提供，以促銷這款新列車組，還有我們何時可以訂TS500。

期待您的回覆。

湯姆・佛德 謹上

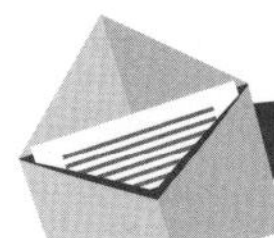

句型這樣替換也可以

1. **Looking forward to your response** 期待您的回覆
2. **We have had numerous requests from...**
   我們有好多來自……的請求
3. **I'll make sure to inform you...** 我一定會通知您……

Unit 03 | 邀請參加記者會

寄信時可以這麼說

Attention Fishermen!

For the third year **in a row** ①, Bragg Creek Fishing Tackle will be doing a **product launch** at this year's show. This will be the most important event for us throughout the year, so we really hope to see you here ②. If you have not **secured** an invitation to the show yet, please contact us as soon as possible ③ so we can get one in the mail to you. Please respond today ④ to book your place at this **fantastic event**.

Look forward to meeting one and all.

Lisa Gamebers

各位釣客們注意囉！

連續第三年①，布拉格溪釣具將在今年的展覽會上作一項**產品上市發表**。這會是我們今年中最盛大的節目，所以我們非常希望能在當天見到您②。如果您還沒**拿到**展覽會的邀請函，請盡早③與我們聯繫，好讓我們寄給您。 請於今日④回覆，預約您在這次**精彩盛會**的席次吧！

期待見到各位！

莉莎・甘柏斯

換句話說

① **the third consecutive year**

→片語in a row表示「連續地；接連地」，使用上通常會置於時間之後，例如：He has won the award for four year in a row.（他連續四年得了獎）。除此之外，另一種表示方式可以用形容詞consecutive，表示「連貫的；無間斷的」，使用上則為「序數詞+consecutive+時間單位」，因此此句可以替換為the third consecutive year。

② **to share the great moment with you 和您共享美好的片刻**

→郵件中，此句的目的為說服收件者一同餐與盛會，除了原句的hope to see you here（希望在當天見到您）以外，也可以用替代句hope to share the great moment with you，表示共享此次盛會的希望。

③ **at your earliest convenience 在您最方便的時候**

→通常在書信往來中表示「盡早回覆」，一般會直覺性的想到as soon as possible，不過，想要再更委婉一點以表示禮貌的話，替代句at your earliest convenience會是更好的選擇。

④ **right now 現在／straight away 馬上／without delay 立刻**

→要催促他人做某事時，可以使用替代片語right now（現在）、straight away（馬上）和without delay（立刻）等。

單字片語急救包

♥ **product launch** *n.* 產品上市發表
♥ **in a row** 連續地；接續地
♥ **secured** *v.* 獲得
♥ **fantastic** *adj.* 極好了；很棒的
♥ **event** *n.* 事件；盛會

回覆時可以這麼說

Lisa,

Thank you for the **notification**. I am crazy about fishing and have been purchasing your products for years ①, so I would very much like to ② attend the show. Is it possible to send me the details ③ by email, or is there a **website** where I can register online ④? If not, can you please mail the invitation to the following address:

Borat Howards 1450 45th Ave Dawsons Ridge 45930

Thanks.

Looking forward to seeing you at the show.

Borat

莉莎：

謝謝您的**通知**。我很狂熱於釣魚，也向貴公司購買產品多年了①，因此我非常想②參加展覽會。能不能請您將詳細內容③以電子郵件寄給我，或是有沒有**網站**可以讓我線上報名④？如果沒有的話，麻煩你將邀請函寄到下列地址：

收件人：柏拉特．霍爾茲 45930道森山市第四十五大道 1450號

謝謝您。

期待在展覽會上看到您。

博拉特

換句話說

① **since your company was founded 自從貴公司創立以來**
→郵件中，此處欲表示寄信者對於此活動的熱情，因此說明他已向公司購買產品多年了。片語for years表示「很多年」，此外，還可以用替代句since your company was founded表示「自從貴公司創立以來（就開始購買）」。

② **will definitely / certainly / surely / absolutely 絕對會**
→郵件中的I would very much like to attend...說明寄信者對於參與活動的高度期望，但是並沒有明確地給予參不參加的直接答覆，而若是使用替代句則為肯定會出席的說法。

③ **timetable / schedule 時間表**
→通常在一場活動中，還會有各個時段的活動細項，因此此處除了可以向對方詢問細節detail以外，也可以要求對方告知活動的時間表，也就是替代詞timetaable或schedule。值得注意的是，同樣為「行程」的兩個字agenda和itinerary分別特指「會議行程」和「會議計畫」，較不適用於此。

④ **book in online 線上登記／get more information about the show 得到更多關於展覽會的資訊**
→郵件中的register online表示「線上報名」，除此之外，也可以用替代句book in online或是get more information about the show，分別表示「線上登記」和「得到更多關於展覽會的資訊」。

單字片語急救包

♥ **notification** *n.* 通知
♥ **website** *n.* 網站

馬上來練習吧！

想要邀請參加記者會時該怎麼說呢？

1. 開頭稱謂　Dear Frank Steely,

2. 問候句

You are invited to ______________________.

______________________________________.

______________________________________.

3. 信件主要內容

______________________________ is attached.

Please __________________________________

____

____

_____________________.

4. 結尾問候句

We ____________________________________

____

_____________________.

5. 署名

Best Regards,

Don Gordon

參考解答 Answer

Dear Frank Steely,

You are invited to the news conference of City Contemporary Art Exhibition on March 20, 2009. As you know, this is the biggest event of our gallery this year, and we would like to have your attendance at the conference. The invitation and the return slip for the conference is attached. Please reply to us before March 5.

We are looking forward to seeing you.

Best Regards,

Don Gordon

中譯

親愛的法蘭克・史迪力：

您已受邀參加二〇〇九年三月二十日的城市當代藝術展的新聞記者會。如您所知，這是本畫廊今年最大的盛事，我們很希望您能出席。邀請函以及回函均已附上。請在三月五日之前給我們答覆。

期待能夠見到您。

唐・葛登 謹上

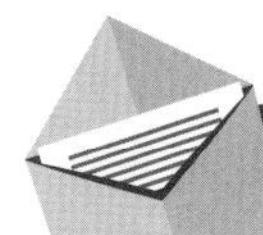

句型這樣替換也可以

1. **If you...please contact us as soon as possible**
   如果您……請儘快與我們聯繫
2. **Thank you for the notification** 謝謝您的通知
3. **I would very much like to attend...** 我非常想參加……
4. **Is it possible to send me...** 能不能寄給我……
5. **Can you please mail the invitation to...**
   可不可以請你將邀請函寄到……
6. **We are well aware of...** 我們非常瞭解……

Unit 04 | 通知違反合約

寄信時可以這麼說

To Jonathan Short,

This letter serves to inform you that you are **in breach of contract**. Therein you clearly **committed** to no releasing any of the details **pertaining to** ① any of the technical aspects ② of the new **vacuum cleaner** until after the launch. The article you wrote ③ last week is not only in breach of that contract, but contains ④ **negative comments**. Unless a formal and public **retraction** and other compensation are made, we will **have no choice but to** take **legal action**.

Yours Sincerely,

Alice Bean

致強納森・休特：

這封信是要通知您，您已經**違約**了。在那份合約裡，您清楚地**保證**在產品發表之前，不會釋出任何**有關**①新款**吸塵器**在技術方面的②細節。但是上週您寫的那篇文章③，不僅違反了那分合約，同時還包含④了**負面的評論**。除非您正式且公開地**撤銷**您的發文，並做出賠償，否則我們將**不得不**採取**法律行動**。

艾利絲・賓恩 謹上

換句話說

① **regarding 有關／with reference to 提及……之事／on the subject of 有關……的主題**

→郵件中的片語pertain to sth.表示「與……有關」之意，除此之外，還可以用替代片語表示：regarding、with reference to和on the subject of。

② **functional aspects 功能方面的／design feature 設計特色**

→一項產品的資訊細節，除了包含郵件中提及的「技術相關層面」technical aspects以外，還有functional aspects「功能方面的」，和design feature「設計特色」等，也可以在此處做替換。

③ **the information you uploaded 你上傳的資訊／the message you posted 你發布的訊息**

→同樣是the sth. you V.的句型，此處也可以用你上傳的資訊the information you uploaded或你發布的訊息the message you posted替換。

④ **includes**

→郵件中的contain一詞表示「包含」，也可以使用字義相同的include表示。

單字片語急救包

- ♥ **commit** *v.* 承諾；保證
- ♥ **negative comment** *n.* 負面評論
- ♥ **vacuum cleaner** *n.* 吸塵器
- ♥ **retraction** *n.* 取消；撤回
- ♥ **legal action** *n.* 法律行動
- ♥ **in breach of contract** 違約；違反
- ♥ **pertaining to sth.** 與……有關；涉及
- ♥ **have no choice but to** 不得不……

回覆時可以這麼說

Dear Alice,

Not only will I not retract any of my **statements** as regards to your product, but I will further continue to release more. ① I believe that not correcting the issues which I have already drawn your attention to many times ② has **forced** me to **make** my findings **public** ③. It is fine if you want to **resort to** legal action, for **at least** this problem as regards your product will be drawn to the public's attention ④at that time. We will see who is the winner then.

Yours Faithfully,

Jonathan Short

艾利絲：

我不僅不會收回我有關你們產品的任何**聲明**，我還會再繼續釋出更多訊息。①我認為你們沒有改正我已經提醒你們好幾次②的問題，**迫使**我將我的發現**公諸於世**③。你想**訴諸**法律行動也沒關係，**至少**屆時有關你們產品的問題就會受到大眾的注意④。我們會知道誰才是贏家。

強納森・休特 謹上

換句話說

① **Neither will I retract..., nor will I stop releasing more.**

我不僅不會撤回……，也不會停止釋出更多訊息。

→郵件中的not only...but (also)表示「不但……而且」，由於郵件中的內容前句說明「不但不會……」，所以也可以將句型改為neither...nor，表示「既不會……也不會……」。值得注意的是，由於原句的後半部分為「我還會……」，使用nor連接時，要在will後面加上stop Ving，以反面說法表示。

② **long time ago 很久以前**

→郵件中，此處欲表示「提醒很多次的問題」，以many times修飾說明I have already draw your attention to。此外，還可以使用替代句long time ago表示「很久以前（提醒的問題）」。

③ **reveal my findings**

→郵件中的片語make sth. public表示「將某事公諸於世」，此外也可以使用動詞reveal表示「揭露」，也就是將此句改為reveal my findings。

④ **notorious**

→郵件中，此處欲表示「法律行動會讓大眾知道產品的問題」，此處也可以單用形容詞notorious表示「惡名昭彰的」。

♥ **statement** *n.* 聲明
♥ **force** *v.* 強迫；迫使
♥ **make public** 將……公諸於世
♥ **resort to** 訴諸……
♥ **at least** 至少

馬上來練習吧！

想要通知違反合約時該怎麼麼說呢？

1.	開頭稱謂	Dear Walter Bolt,
2.	問候句	We have ______________________________ ______________________________ ____________________.
3.	信件主要內容	Consider this a final warning. If we __________ ______________________________ ______________________________ ______________________________ ____________________.
4.	結尾問候句	Regards,
5.	署名	Patrick Gibbon

參考解答 Answer

Dear Walter Bolt,

We have received numerous complaints from end customers that you have been selling our product for well above the maximum allowed retail price as per our contract. We have confirmed this by sending a staff member to your store. Consider this a final warning. If we hear of this again, we will be forced to withdraw our support and products from your outlet.

Regards,

Patrick Gibbon

中譯

致華德・波特：

我們已接獲許多客戶投訴，您以遠遠超過依照我們合約所允許的最高零售價銷售我們的商品。我們已經派人到您店裡確認過。這將是最終警告。若我們再次聽聞此事，將不得不撤回我們的支持以及您賣場的商品。

派崔克・吉本 謹上

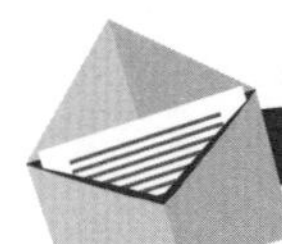

句型這樣替換也可以

1. **This letter serves to inform you that...** 這封信是要通知您……
2. **Unless...we will have no choice but to...**
   除非……，否則我們只好……
3. **It is fine if you want to resort to...** 你想訴諸……也沒關係

Unit 05 詢問違反合約通知

寄信時可以這麼說

To Legal Department,

I was recently **served with** a notice to pay out the remaining period on my contract due to **non-fulfillment** of its **conditions**. I must admit ①, it never **crossed my mind** ② that this would occur. I have moved from where I had your cable services ③ and am unable to purchase them in my new location. I never had any **issue** with the service, but am merely now unable to use the remaining year on the contract due to the move. Please help me with ④ this matter.

Yours Sincerely,

Gina Black

致法律事務部：

我最近**接到傳票**，要我支付合約剩餘期間**未履行**其**條件**的費用。我必須承認①，我從沒**想過**②會發生這種事情。我已經搬離使用你們的有線電視服務③的區域，而且目前我的新居處並無法購買你們的服務。我對你們的服務從來沒有過**異議**，只是現在我因為搬遷而無法使用剩下的一年合約。麻煩請幫助④我解決這個問題。

吉娜．布萊克 謹上

換句話說

① **Admittedly 不可否認地**

→郵件中的I must admit表示「我必須承認」，而替代詞admittedly表示「不可否認」，可以在此作替換。值得注意的是，通常admittedly會使用於「不情願地承認」之情況下，例如：Admittedly, there is a serious problem that must be dealt with here.（不可否認，現在有緊急要處理的問題）。

② **occurred to me**

→郵件中的cross one's mind表示「（某人）突然想起」，相似地片語還有替換句地occurred to sb.。值得注意的是，兩種片語地主詞皆為「想起的事物」，並非「想起事物的人」。

③ **internet services 網路服務／**
music streaming services 音樂串流服務

→現今的社會中，「服務」的種類包羅萬象，除了郵件中提及的cable sevice有線電視服務以外，還有網路服務internet service和音樂串流服務music streaming service等，可以在此作替換。

④ **to solve / to deal with**

→help一詞的用法除了郵件中的help sb. with sth.以外，還有help sb. to V.的形式。因此此處可以替換為help me to solve this matter.或是help me to deal with this matter.。

- ♥ **non-fulfillment** *n.* 未完成
- ♥ **condition** *n.* 條件
- ♥ **issue** *n.* 問題；議題
- ♥ **serve with** 送達（傳票等）
- ♥ **crossed one's mind** 突然想起

回覆時可以這麼說

Dear Gina,

Thank you for the email. I am sure we can **come to terms** on this. Obviously ① when we saw your **cancellation**, we did not know the details and sent out a pre-formatted email. I apologize if it was not expected ②. We do have a means by which you can avoid paying the full **penalty** on our website. An administrative charge ③ of $50 will be required to complete this, but once we have the needed documents, you will no longer **be liable for** the contract ④.

Yours Sincerely,

Bob White

親愛的吉娜：

感謝您的來函。我相信我們在這件事上能**達到協議**。顯然①，當我們看到您的**取消**通知時，在不知道細節的情況下發出了預先格式化的電子郵件。如果讓您感到意外②，我很抱歉。我們的網站確實有讓您避免支付**罰款**的方法。完成解約動作將需要行政費用③50元，一旦我們收到所需文件，您就不需要再**負合約上的責任**④。

鮑伯・懷特 謹上

換句話說

① Apparently

→郵件中的obviously表示「顯然地；顯而易見地」，同義單字apparently也可以在此作替換。

② beyond your expectation

→expect一詞表示「預料；期待」，因此郵件中的is not expected表示「某事（收到傳票）是某人沒有預料到的」，也就是「出乎預料」，也可以用beyond expectation表示。值得注意的是，beyond在此表示「令……無法理解；非……所能及」，相似用法還有beyond words（無法用言語表達）。

③ A surcharge 額外收費／A commission 手續費

→一般情況下，辦理事務所需要支付的費用會用行政費用administrative charge、surcharge額外收費，或是commission手續費表示，因此此處可以互相做替換。

④ not be liable for the contract any more

→No longer和not any longer均意為「不再」，一般視之為同義詞，但值得注意的是，no longer相較之下是比較正式的用法。

單字片語急救包

- ♥ **cancellation** *n.* 取消
- ♥ **penalty** *n.* 罰款；處罰
- ♥ **come to terms** 達成協議
- ♥ **be liable for** 具有法律責任的；有義務的

馬上來練習吧！

想要回應售價與產品內容時該怎麼說呢？

1. 開頭稱謂　Dear Patrick,

2. 問候句　I think ______________________________,
______________________________,
______________________________,
____________________.

3. 信件主要內容　______________________________.
Hence ______________________________
____
____________________.

4. 結尾問候句　Thank you for your continued support.

5. 署名　Walter Bolt

參考解答 Answer

Dear Patrick,

I think there has been a misunderstanding as to regards my pricing. While it may appear that I am selling above your recommended value, I bundle your product with the needed cables, batteries and memory card. As you know your product comes without these. Hence I feel that I am in fact meeting a customer need by supplying the product in this way.

Thank you for your continued support.

Walter Bolt

中譯

親愛的派崔克：

我想您對於我的定價有所誤解。雖然看起來我似乎以比您所推薦的售價要高的價格來販售，但是我將您的產品與所需電線、電池以及記憶卡整包販售；而您的商品原來是沒有這些的。因此，我覺得我這樣的供貨方式事實上是滿足客戶需要的。

懇請繼續支持，不勝感激。

華德．波特

句型這樣替換也可以

1. **I was recently served with a notice of...** 我最近接到……的傳票
2. **I am sure we can come to terms on...**
   我確信我們能在……上達成協議
3. **I apologize if it was not expected** 如果讓您意外，我感到很抱歉

寄信時可以這麼說

To **Billing Department**,

We recently purchased a new living room set ① from your store and signed on for the **no interest, no payments** plan ②. I was informed at the time that all the **relevant** details as to ③ when payment would be needed ④ would be **forwarded** to me by mail. As yet I have not received anything. I want to make sure if there was any mistake.

Regards,

Edward White

致**帳務部**：

我們最近在你們店裡購買了**一組**新的**客廳傢俱**①，並簽了「**免利息免付款**」的方案②。當時我被告知所有**相關**細節，也就是③何時需要付款④，會以郵件的方式**轉寄**給我，但是目前為止我還沒有收到任何東西。我想確認這之中是否有什麼錯誤。

愛德華．懷特 謹上

換句話說

① **leather sofa set 皮沙發組／bedroom set 臥室傢俱組／bed set 床組**

→除了郵件中的客廳傢俱living room set以外，還可以注意其他種傢俱的名稱寫法：皮沙發組leather sofa set、臥室傢俱組bedroom set和床組bed set，也可以在此作替換使用。

② **installment plan 分期付款方案**

→常見的付費方案除了郵件中提到的「免利息、免付款方案」以外，還有分期付款方案installment plan，可以在此作替換使用。

③ **regarding**

→值得注意的是，雖然as to和as for都可表示「至於」，但是這裡as to就如同regarding，是「關於」的意思；而as for則用於句首，不適用於此。

④ **how much should I pay for the down payment 我該付多少的頭期款**

→承第③點，選擇分期付款installment plan就一定會談及頭期款down payment，因此此處可以替換為how much should I pay for the down payment.（我該付多少的頭期款）。

單字片語急救包

- ♥ **Billing Department** *n.* 帳務處理處（部）
- ♥ **relevant** *adj.* 相關的
- ♥ **forward** *v.* 轉寄
- ♥ **no interest, no payments** 免利息、免付款

回覆時可以這麼說

Dear Edward,

Thank you for **doing business with** us. We hope you are pleased with ① your **purchases**. I have **investigated** your **account** and it appears that ② the relevant **documentation** was sent to you the day after ③ your purchase. Now that you haven't received it yet, I have attached another copy of this for your records ④. Would you be so kind as to reply to this email to confirm your address for our files?

Thank you very much.

Sincerely,

Cheryl Bird

親愛的愛德華：

感謝您**與**我們**交易**。希望您滿意①您**購買的商品**。我已**查**過您的**帳戶**，看來②相關**文件**已經在您購買後的那一天③發送給您了。由於您還沒收到，我已附上另一個副本供您留存④，麻煩您回覆此郵件，以確認您的地址，讓我們歸檔。

非常感謝您！

雪柔．博德 謹上

換句話說

① **are satisfied with**

→表示「對……感到滿意」除了可以用郵件中的be pleased with以外，還可以用替換句be satisfied with表示。

② **it seems that 似乎**

→郵件中的it appear that表示「看起來好像；似乎」，另一種常用的片語為it seems that有時也會用it seems to sb. that表示，例如：It seems to me (that) he isn't the right person for the job.（在我看來，他不是這項工作的合適人選）。

③ **right after 馬上／two days after 兩天後**

→郵件中的時間補語the day after修飾後面的your purchase，此處也可以視情況替換為right after「馬上」，或two days after「兩天後」。

④ **for your reference 供您參考**

→郵件中的for your record表示「供您留存」，相似用法還有for your reference表示「供您參考」，雖然在意義上並不完全相同，不過用於此處沒有太大的差別。

♥ **purchase** *n.* 所購買之物
♥ **investigate** *v.* 調查
♥ **documentation** *n.* 文件（總稱）
♥ **account** *n.* 帳戶
♥ **do business with**
和……做生意、做買賣

馬上來練習吧！

回覆付款通知時該怎麼說呢？

1. 開頭稱謂　Dear Billing department,

2. 問候句　Thank you for ______________________________.
______________________________________.

3. 信件主要內容　While ______________________________
____
______________________________________.
Could you ______________________________
____
____________________?

4. 結尾問候句　Thanks in advance.

5. 署名　Allen Joyce

參考解答 Answer

Dear Billing department,

Thank you for sending me the reminder. I also appreciate that chance to save a little bit on my premium. While I understand that doing things online must be cheaper, I am not very good with computers and have been unable to complete the payment. Could you let me know how else I could qualify for the discount if I can't complete it online?

Thanks in advance.

Allen Joyce

中譯

致帳務部：

謝謝您寄發提醒函給我，我同時也感謝有機會能省下一點保費。雖然我明白在線上辦理事宜會比較便宜，但是我對電腦不是很在行，而無法完成付款。能不能告訴我其他方式，讓我不需要在線上完成付款也能享有優惠呢？

在此先謝謝您。

艾倫．喬埃思

句型這樣替換也可以

1. **We have sent out this early reminder that...** 我們已經寄上提醒函……
2. **Thank you for the timely reminder** 感謝您的及時提醒
3. **I have been very happy with...** 我一直以來都對……很滿意
4. **Thank you very much for offering us...** 非常感謝您為我們提供……
5. **We recently purchased...from...** 我們最近從……購買了……
6. **We hope you will be satisfied with...** 希望您會對……滿意

Unit 02 尚未入款通知

寄信時可以這麼說

Dear Valued Customer,

This notice serves as a reminder ① that your invoice 159684 for $12300 is still **outstanding** ②. You still have 7 days left from the date of this email ③ to pay off the outstanding **balance in full**. Thereafter ④ **late payment penalties** will be **in effect**.

Thank you for your prompt attention to this matter.

Regards,

Billing and Invoicing

尊貴的顧客您好：

這則通知是要提醒您①，您金額為12300美元，號碼為159684的收據，至今仍**未繳納**②。自此封郵件日期算起③，您還有七天可以付清未支付的**剩餘款項全額**④。在此之後，將會**產生逾期繳款的罰金**。

若您能立即注意此問題，將不勝感激。

帳務收發部 謹上

換句話說

① **to remind you**

→郵件中的as a reminder表示「作為提醒」，也可以轉換詞性，使用替換句serves as to remind you。

② **remains unpaid**

→郵件中的outstanding在此譯作「未支付的」，而非「傑出的」，因此也可以用替代句remains unpaid表示那筆錢目前的狀態是「仍未付清的」。

③ **since this email is sent**

→郵件中的from the date of this email表示「從郵件的日期算起」，另外也可以用替代句since this email is sent，表示「自這封郵件寄送之時」。

④ **After that**

→替換詞after that和thereafter同樣表示「自此之後」，that是代替提一句提到「七天緩衝期間」的代名詞，用以避免太多重複的字詞出現。

♥ **outstanding** *adj.* 未解決的；未支付的
♥ **balance** *n.* 剩餘部分；餘款
♥ **late payment penalty** *n.* 逾期付款罰金
♥ **in full** 全額；全部
♥ **in effect** 生效；實行

回覆時可以這麼說

Dear Billing Department,

I am a little surprised to have received your e-mail. ① We have **discontinued** ② service with your company two months ago, ③ yet continue to receive invoices. Please update ④ your records. Our account was **in good standing** when all services with your company were cancelled. I have an **original letter** of cancellation which I can forward to you if you cannot obtain a copy from your files.

Regards,

Michael Mannen

致帳務部：

收到您的郵件讓我感到有點意外。①我們在兩個月前③已經**終止**②你們公司的服務，卻仍繼續收到發票。請更新④你的記錄。我們的帳戶在取消貴公司所有的服務時是**有良好信譽的**。我這裡有取消服務的**原始信件**，如果您無法從你的檔案中找到副本，那麼我可以轉寄給您。

麥可．曼恩 謹上

換句話說

① I didn't expect to receive your e-mail.

我沒想到會收到您的郵件。

→郵件中的首句表示「收到您的郵件讓我感到有點意外」，換言之，就是「我沒想到會收到您的郵件」之意，所以可以用替代句I didn't expect to receive your e-mail.表示。

② cancelled / terminated

→郵件中的動詞discontinued由字根continue（繼續）加上字首dis組成，字義為「終止；中斷」，同義的單字還有cancel和terminate可以在此作替換使用。

③ half a year ago 半年前／a few months ago 幾個月前

→在英文中，通常表示時間的補語會置於句尾，也就是原句的two months ago，除此之外，其他時間表達方式如：半年前half a year ago或是幾個月前a few months ago也可以在此作替換使用。

④ renew

→同樣能表示「更新」的動詞還有renew，可以在此作替換使用。值得注意的是， renew常用於表示「續約」，例如：renew the subscription和renew the contract等。

- ♥ **discontinue** *v.* 停止；中斷
- ♥ **original letter** *n.* 原始信件
- ♥ **in good standing** 名聲良好

馬上來練習吧！

想要表示尚未入款通知時該怎麼説呢？

1. 開頭稱謂　To Honest Joe's Car Repair,

2. 問候句　We ______________________________

______________________________.

3. 信件主要內容

____

______________________________.

Please remit this to us ______________________________

____

____

______________________________.

4. 結尾問候句　Yours Sincerely,

5. 署名　Billing and Collections Department

參考解答 *Answer*

To Honest Joe's Car Repair,

We have not yet received payment for the past two invoices we have sent to you. The total outstanding amount is now $378.12. Please remit this to us within the next two weeks or we will be forced to terminate your telephone service.

Yours Sincerely,

Billing and Collections Department

中譯

致誠信的喬汽車修繕公司：

我們還沒有收到我們已經寄給您的過去兩張收據的款項。目前未清償的款項總金額為 378.12美元。請在未來兩週之內完成匯款，否則我們將不得不中斷您的電信服務。

帳務部 謹上

句型這樣替換也可以

1. **This notice serves as a reminder that...**
   這則通知是要提醒您……
2. **You still have 7 days left from...** 自……算起，您還有七天時間
3. **I am a little surprised to have received...**
   收到……我感到有點意外

Unit 03 詢問存款事宜

寄信時可以這麼說

To Billing,

My name is Andy Smith, account number 569483. I sent a **check** to your company three weeks ago for payment of ① my invoice 203302. Since then ②, I have been watching my account, and the check **in question** ③ has still not ④ been **deposited**. Can you please confirm receipt of the check for $134.76?

Thank you.

Andy Smith

致帳務部：

我的名字是安迪・史密斯，帳戶號碼為 569483。我在三個星期前寄了一張**支票**到貴公司以支付①我那張號碼為203302的發票。從那時起②，我一直查看我的帳目，**我說的**③那張支票至今仍然還沒④**存進去**。可否請您確認是否收到134.76美元的支票呢？

謝謝您。

安迪・史密斯

換句話說

① to pay for

→郵件中的for payment of表示「以支付……」，在此也可以轉換名詞payment為動詞pay，改寫成替換句to pay for。值得注意的是，payment和pay後分別接的介系詞為of和for，不可隨意替換。

② From then on

→郵件中的since表示「自從（某時起）」，意義等同於from一詞，所以此句也可以用替換句改寫為from then on。值得注意的是，若要使用since，後面不須加上on；若要使用from，則要加上。

③ I've mentioned 我所提及的

→片語in question表示「正在討論的」，而另一種要表達「先前提及的話題」的方法還可以用I have mentioned或是I just mentioned。

④ not yet

→表示「尚未」除了使用郵件中的still not...以外，還可以用not yet...，因此此句可以改寫為the check in question has not yet been deposited.。

♥ **check** *n.* 支票
♥ **deposit** *v.* 存入
♥ **in question** 討論中的

回覆時可以這麼說

Dear Andy Smith,

Thank you for your email. We have **indeed** received the check in question. Typically ① we do **deposits within** the same week of having received checks. However, our accounts **receivable** clerk ② had been on vacation ③ during the past three weeks and there is thus a **backlog** ④. We expect that the check will be deposited in the next couple of days.

Regards,

Grace Bullock

親愛的安迪．史密斯：

謝謝您的來函。我們**的確**已經收到您說的那張支票。通常①我們會**在**收到支票的同一個星期**內**就處理**存款**。然而，我們的處理帳戶**應收帳款的**行員②過去三週去度假了③，因此才會有**積壓**的情形發生④。我們預計支票將會在未來幾天內就存入。

葛瑞斯．布拉克 謹上

換句話說

① Normally / Commonly

→郵件中的typically表示「通常」，同義的副詞還有normally和commonly也可以在此作替換使用。值得注意的是，這類的副詞除了置於句首以外，也可以置於主詞之後、動詞之前，例如：we normally have dinner at 7.（我們通常在七點吃晚餐）。

② staff / operator

→郵件中的clerk就是「員工；行員」之意，因此可以用staff「員工」或是operator「操作員；業務員」替換表示。

③ on sick leave 請病假／hospitalized 住院治療

→be on a vacation在英文中為固定用法，表示「在度假；休假中」，除此之外，現實情況中還可能遇到的情形還有：請病假be on sick leave和住院治療hospitalized，在此處可以做替換使用。

④ therefore caused a delay 因此造成延遲

→郵件中的backlog一詞表示「積壓」，也就是原先應該要完成的作業不斷累積而成的工作，因此此處也可以改以「延遲」解釋，也就是替換句的therefore caused a delay。

♥ **indeed** *adv.* 確實；實在
♥ **deposit** *n.* 存款
♥ **receivable** *n.* 應收帳號
♥ **backlog** *n.* 積壓
♥ **within...** 在……之內

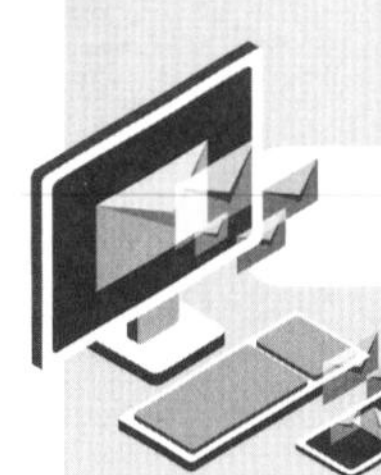

馬上來練習吧！

想要詢問存款事宜時該怎麼說呢？

1. 開頭稱謂　To Amce Telephone Services,

2. 問候句　Please find ______________________________
______________________________.

3. 信件主要內容　I apologize ______________________________
______________________________.
____
____
______________________________.

4. 結尾問候句　Thank you for been so understanding.

5. 署名　Regards,
Honest Joe

參考解答 Answer

To Amce Telephone Services,

Please find the enclosed check for the outstanding amount that I owe to you. I apologize for not paying this on time the past couple months. We have been understaffed and I have not been able to keep up with all the management work.

Thank you for been so understanding.

Regards,

Honest Joe

中譯

致AMCE電信服務：

附上一張我應該給您的未清償款項支票。很抱歉過去幾個月沒有按時繳納。我們一直處在人手不足的狀態，我還沒能跟上所有的管理工作進度。

感謝您的體諒。

誠信的喬 謹上

句型這樣替換也可以

1. **Please update your records** 請更新你的記錄
2. **Can you please confirm...** 可否請您確認……呢
3. **We expect that the check will be deposited in...**
   我們希望支票將會在……存入

Unit 04 開立發票

寄信時可以這麼說

To Accounts Payable,

Western Car Sales, Last month our invoice to your company was **returned to** us by the postal service ① as address no longer **valid** ②. Can you please **provide** your updated contact details ③ **for** our Billing Department ④ to send to? If you can provide contact names, fax and telephone numbers as well as a **physical address**, it would be greatly appreciated.

Yours Sincerely,

Ronald Douglas

致西方汽車銷售應付帳款部：

上個月我們寄到貴公司的發票，因為地址**失效**②而被郵局①**退回**。可否請您**提供**更新後的聯絡細節③**給**我們的帳務部門④，好讓我們寄發票給您？若您能提供聯絡姓名、傳真、電話號碼以及**收信地址**，將不勝感激。

羅納德．道格拉斯 謹上

換句話說

① **the express delivery 快遞／Federal Express 聯邦快遞**

→除了郵件中提及的郵局以外，還有別種物流運送的種類，例如替換詞快遞the express delivery和聯邦快遞Federal Express，皆可以在此處做替換。

② **because of the invalid address**

→郵件中原句的as在此用以表示原因，而與原因密不可分的連接詞because也可以在此直接替換，另外，也可以用because of +N片語表示，因此此句可以改寫為...because of the invalid address。

③ **information**

→通常提及「聯絡資訊」在英文常以contact information表示。值得注意的是，information一詞為不可數名詞，因此不能在其後加上複數s。

④ **General Accounting Department 財務部／accountant 會計人員**

→公司內部與帳務相關的部門還有財務部General Accounting Department和會計人員accounting，在寫法上可以注意一下。

單字片語急救包

♥ **valid** *adj.* 有效的
♥ **physical address** *n.* 實體地址
♥ **returned to** 退回（給某人／某地）
♥ **provide sth. for sb.** 提供……給……

回覆時可以這麼說

Dear Ronald,

My apology if our change of address ① notifications did not **reach** you ②. We have **moved** location ③ **to** a bigger **facility**. The new location is

4503 12th Ave Small Ville Utah 30333

You can reach me personally ④ at 390905-4859, or fax me at 390-905-4839

Regards,

Rhonda Jones

親愛的羅諾：

若您未**收到**我們更改地址①的通知②，我在此致歉。 我們已經**遷移至**③另一個較大的**地點**，新的地址是：

30333 猶他州 斯莫爾鎮 十二大道 4503號

您可以打390-905-4859這支電話號碼與我本人④取得聯繫，或傳真至390-905-4839 給我。

藍達．瓊斯 謹上

換句話說

① **contact details 聯絡資料／contact number 聯絡電話**

→通訊方式除了郵件中提及的住址address以外，還有替換詞聯絡資料contact details和聯絡電話contact number，在此可以替換座使用。

② **you didn't get informed of our changed address**

→郵件中的原句是以被動的方式表示「我們更改地址的通知沒有被您收到」，而替換句子you didn't get informed of our changed address則表示「你沒有收到我們更改的地址」。

③ **relocated**

→郵件中的moved location表示「移動位置」，可以直接以動詞relocate替換。因此此句可改為We have relocated to a bigger facility.。

④ **directly 直接地**

→郵件中的personally表示「個人地；親自」，換言之，就是「直接」與本人聯絡，因此在此可以替換為副詞directly。

單字片語急救包

♥ **reach** *v.* 抵達；到達；與……取得聯繫
♥ **moved to...** 搬移至……
♥ **facility** *n.* 場所

馬上來練習吧！

想要開立發票時該怎麼說呢？

1. 開頭稱謂　To the Billing Department,

2. 問候句　____________. In that time we ____________, ____________. We are about to ____________ ____________________________.

3. 信件主要內容　Please ________________________
____
____
____
________________________.

4. 結尾問候句　Thanks.

5. 署名　Claire Lee in Accounts Payable

參考解答 *Answer*

To the Billing Department,

For the past seven months your company has received direct deposits from our company for your services. In that time we have only received 3 receipts, one of them a duplicate. We are about to undergo our 4 year external audit and really need the copies of all the invoices in order to show the accountants.

Please have them sent to us as soon as possible.

Thanks.

Claire Lee in Accounts Payable

中譯

致帳務部：

過去7個月，貴公司都從我們公司直接收取訂金。這段時間，我們只收到三張收據，其中一張還是重複的。我們即將進行四年的外部審計，真的需要所有的收據副本以示會計人員。

請盡快將它們寄過來。

謝謝。

應付帳款部 克萊爾・李

句型這樣替換也可以

1. **Can you please provide...** 您能提供……嗎？
2. **If you...it would be greatly appreciated** 如果您……，將不勝感激
3. **My apology if...** 若……我在此致歉

Unit 05 要求寄送發票

寄信時可以這麼說

To the Billing Department,

I recently **paid** my account **up** in full with your company. However, I have not received any receipt from your company yet. I will need a receipt for **tax purposes** ① which is almost near the **deadline**. Can you please ② either resend the receipt to me, or send me an email with the receipt attached ③? And please **note** that this is really an urgent need ④, so it would be nice if you can reply me as soon as possible.

Thanks in advance.

Bryan Grey

致帳務部：

我最近已全額**付清**與貴公司的帳款，卻還沒收到貴公司的收據。我需要收據**做報稅之用**①，**截止日期**也快到了。能否請您②重新寄收據給我，或是寄給我一封附加收據的電子郵件③？並請您**注意**，這真的是一個緊急的需求④，如果您能盡快回覆我會很感激。

在此先向您致謝。

布萊恩．葛瑞

換句話說

① **for payment request 做請款之用／for accounting records 做會計帳目紀錄之用**

→郵件中的purpose表示「目的」，因此for sth. purpose表示「以……目的（做某事）」。除此之外，還可以使用替代句for payment request做請款之用，或for accounting records做會計帳目紀錄之用等。

② **Would you mind to...? 你介意……嗎？**

→通常在英文的使用，要表達請求時常會使用can作為開頭，例如郵件中的Can you please ...，不過若要更有禮貌的提出要求時，可以更改為替換句的方式Would you mind...開頭，表示「你介意……嗎？」。

③ **attach the receipt with an email**

→電子郵件常會使用附加檔案，而其表現方式主要有動詞attach和名詞attachment兩種。其中，attach又可以分為主動或被動，主動為替代句的用法attach sth. with an email，而被動方式則為郵件中的原句send an email with sth. attached。

④ **we need this urgently**

→郵件中的urgent need作為名詞，表示收據是「緊急需要的東西」，而若是改為替代句we need this urgently則是用urgent 的副詞形式urgently修飾「需要的迫切感」。

♥ **deadline** *n.* 截止日期
♥ **tax** *n.* 稅 *v.* 繳稅
♥ **purpose** *n.* 目的；原因
♥ **note** *v.* 注意；留意；筆記
♥ **pay up** 付清

回覆時可以這麼說

Dear Bryan,

Thank you for your email. Our records show that ① we mailed the receipt to you on the 14th of last month ②. I apologize if it did not reach you. Please find a **PDF** copy of receipt attached to this email. ③

Additionally ④, could you please **confirm** your address **with** us to ensure that future receipts do reach you. You can reply to this email with your address.

Thanks.

Billing Department

親愛的布萊恩：

感謝您的來信。我們的記錄顯示①，我們在上個月的14日②將收據寄給您了。如果您沒有收到，那麼我很抱歉。請您找這封郵件所附的PDF格式收據副本。③

此外④，麻煩您**和**我們**確認**地址，以確保將來寄收據時你可以收得到。您可以將您的地址回覆在這封電子郵件裡。

謝謝您。

帳務部 謹上

換句話說

① **According to our record, 根據我們的紀錄**
→郵件中的...show that視為表示根據來源，因此也可以用according to片語表示。值得注意的是，此片語在使用上通常會置於句首，例如：According to Emma, she and jack just broke up yesterday.（根據艾瑪所說，她和傑克在昨天分手了）。

② **two weeks ago 兩週之前／two days after your account was settled 帳單結清後兩天**
→在英文的使用上，通常表示時間的補語或副詞會置於句尾，也就是郵件中原句的the 14th of last month，此外，也可以使用替換句兩週之前two weeks ago或是帳單結清後兩天two days after your account was settled。

③ **Please find the attached PDF copy of receipt in this email**
→郵件中原句的attached在此作為動詞，表示「附件」修飾前面提到的主詞a PDF copy of receipt，而替代句子中的attached則是形容詞，雖然拼法相同，使用上卻有所差異。

④ **Also**
→郵件中的副詞additionally譯作「此外」，暗示即將要提及原本話題以外的事情，因此此處也可以用also替換使用。

♥ **PDF** *n.* 便攜式文件格式（= portable document format）
♥ **confirm sth. with sb.** 向某人確認某事

馬上來練習吧！

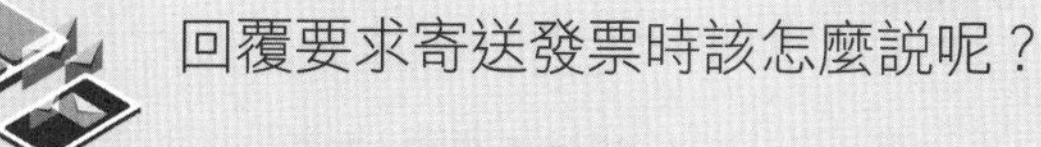

回覆要求寄送發票時該怎麼說呢？

1. 開頭稱謂　Dear Claire,

2. 問候句　Thanks for the email.

3. 信件主要內容

I ______________________________.

Please ______________________________.

____

____

______________.

4. 結尾問候句　Thank you.

5. 署名　Ken Bent

參考解答 Answer

Dear Claire,

Thanks for the email. I have attached a pdf copy of all the invoices and receipts we have for your company on file. Please go through them and confirm that everything is in order. I have also included a copy of next months invoice in case you need it for the audit.

Thank you.

Ken Bent

中譯

親愛的克萊兒：

感謝您的來信。我已經附上一個PDF檔案，此為我們應給貴公司的所有收據和發票。請核對並確認一切都正確。我同時還附上一份往後幾個月的發票，以作為您需要用來審計之用。

謝謝您。

肯・班特

句型這樣替換也可以

1. **You can reach me personally on...** 您可以打……與我取得聯繫
2. **Can you please resend...to me** 可不可以請您重新寄……給我
3. **Please give us a PDF copy of...** 請給我們一份……的 PDF 格式副本

Unit 01 推銷產品

寄信時可以這麼說

Dear Dealers,

Our latest ① catalogue of Brooks skin care ② **product line** has been published ③. There are many new items in this product line that we **carry** this year. **Enclosed** is the PDF file of the catalogue, if you need some more information or test samples, please let us know.

We **highly value** the **partnership** with you over the years, and hope to continuously provide you with the best service ④.

Sincerely,

Mike Rogers

親愛的經銷商們：

我們布魯克斯護膚②**產品系列**的最新①目錄已經發表③了。今年在這個產品系列，我們**推出**了很多新品。**隨信附上**這份目錄的 PDF格式檔案，若您需要更多資訊或是樣品，請告訴我們。

我們**相當重視**多年來與您的**合作關係**，並希望能繼續提供您最佳的服務④。

麥可．羅傑斯 謹上

換句話說

① **newest 最新的／brand new 嶄新的**

→郵件中的lastest表示「最新的」，很常用於英文中，例如：Micheal Jackson's lastest album（麥可傑克森的最新專輯）、Disney's lastest movie（迪士尼最新電影），除此之外，還可以用brand new片語，或是形容詞new的最高級newest表示。

② **KS III cosmetics KS 三代化妝品／Home Chef Cookware 家廚炊具／the household appliances 家電用品**

→郵件中的商品維護膚系列skin care，其他種類的商品可以參考替換詞：KS三代化妝品KS III cosmetics、家廚炊具Home Chef Cookware、家電用品the household appliances等。

③ **out 出版；問世**

→產品的推出除了常用郵件中的publish一詞以外，sth. is out也是常出現的用法，例如：The new movie is out in September.（最新電影在九月上映）。不過相較之下，out的用法較不正式。

④ **maintain this good relation with you 與您維持這良好的關係**

→郵件中的最後一句欲表示寄信者對於雙方關係未來的期許，也就是希望能繼續合作的關係，因此也可以用替代句maintain this good relation with you表示「與您維持這良好的關係」。

單字片語急救包

- ♥ **product line** *n.* 產品系列
- ♥ **carry** *v.* 備有貨品；有……出售
- ♥ **enclose** *v.* 隨信附寄
- ♥ **partnership** *n.* 合作關係
- ♥ **highly value** 相當重視

回覆時可以這麼說

Dear Mike,

The new catalogue **definitely came in time** since many of our customers have been inquiring about the new products.

We would also like a few **hard copies** of the catalogue in order to ① be able to **present** them to the customers. At this stage ②, we only need 5 extra copies. We will also **look into** ③ the new products and **get back to** you once ④ we need more information about them.

Regards,

Jane Steely

親愛的麥可：

這本新目錄**來得正是時候**！我們有好多客戶都在詢問新產品了。

我們想要幾本目錄的**印刷本**，好讓①我們可以**拿給**客戶**看**。目前②我們只需要五本。我們也會**仔細研究**③你們的新產品；一旦④我們需要更多產品資料，就會**再跟**你們**聯絡**。

珍・史迪力 謹上

換句話說

① **so as to**

→郵件中的in order to用以表示做事的「目的」，同樣功能的片語還有替換詞so as to，通常這兩種用法是被視為一樣的，可以互相替換使用。

② **Currently 目前；現在**

→stage一詞除了表示「舞台」以外，還有「階段」之意，所以郵件中的片語at this stage表示「目前的階段」。除此之外，還可以直接用副詞currently替換表示。

③ **check out**

→郵件中的片語look into表示「仔細研究；深入研究」，除此之外，也可以用替換詞check out表示。不過相較之下，後者更常用於英文的口頭對話中。

④ **if / whenever**

→郵件中的此句欲表示「只要我們有需要……就會……」，除此之外，還可以用「如果」if表示，或是「每當」whenever替換使用。

- ♥ **hard copy** *n.* （電腦輸出的）紙本；印刷本
- ♥ **present** *v.* 出示；呈現
- ♥ **definitely came in time** 來的正是時候
- ♥ **look into** 深入研究
- ♥ **get back to sb.** 再與某人聯絡

馬上來練習吧！

回覆推銷產品時該怎麼說呢？

1. 開頭稱謂　To Howard,

2. 問候句　Thanks for ____. ____________________________.

3. 信件主要內容　I have included ____________________________.
If you ____,
____,
please do not hesitate ____.

4. 結尾問候句　Thanks.

5. 署名　Dean Speke

參考解答 Answer

To Howard,

Thanks for taking my call the other day. I hope that the samples we spoke about are now safely at your office and that you have had the chance to look them over. I have included a fair amount of product literature with them. If you have any questions yourself, or just want to find out more information, please do not hesitate to give me a call on my cell.

Thanks.

Dean Speke

中譯

霍爾：

感謝您那天接我的電話。我希望我們談到的樣品現在已經安全地抵達您們公司，讓您有機會仔細檢查。我已經一起附上相當數量的產品資料。

如果您本身有任何問題，或是想知道更多訊息，請儘管撥打我的手機與我連絡。

謝謝。

狄恩・史皮克

句型這樣替換也可以

1. **If you need some more...please let us know**
   若您需要更多……請告訴我們
2. **We highly value the partnership with...**
   我們相當重視與……的合作關係
3. **We would also like a few...** 我們還想要一些……

Unit 02 索取試用品

寄信時可以這麼說

Dear Mike,

Thanks for the new catalogue of Brooks skin care products ①. I did see that there are a few new products in it. Since this is a very popular ② product line, I would like to know more about the details ③ of the products. I **wonder** if we can **set up** ④ a meeting with you for more information and get some test samples. Please let us know when will be convenient for you.

Thank you.

Regards,

Jennifer Anderson

親愛的麥可：

謝謝您寄來布魯克斯護膚品①的新目錄。我的確有看到幾樣新產品。由於這產品系列非常受歡迎②，我希望能夠了解新產品的更多細節③。我**在想**我們是不是能夠**安排**④一個時間，跟你們索取更多資訊和試用品。請讓我們知道您什麼時候方便。

謝謝您。

珍妮佛．安德森 謹上

換句話說

① **toilet detergents 馬桶清潔劑／**
eco-friendly cleansing materials 環保清潔用品／
disposable contact lenses 拋棄式隱形眼鏡

→郵件中的商品維護膚系列skin care，其他種類的商品可以參考替換詞：馬桶清潔劑toilet detergents、環保清潔用品eco-friendly cleansing materials，和拋棄式隱形眼鏡disposable contact lenses等。

② **attractive 吸引人的**

→產品通常除了會用「受歡迎」popular形容之外，常用的還有「吸引人的」attractive，因此在此可以做替換使用。

③ **the features 特色／the functions 功能**

→有關商品的相關資訊，除了可能提到商品細節detail以外，還可能會提及商品特色the feature of the products、產品功能the fuctions of the products等。

④ **arrange**

→英文的商用書信中，通常會談及「安排會面」的事項，除了用郵件中的set up a meeting以外，還可以用動詞arrange替換，所以此句可以改用arrange a meeting表示。

♥ **wonder** *v.* 想知道；疑惑

♥ **set up** 建立；安排

回覆時可以這麼說

Hi Jennifer,

I **am glad to hear back from** you ①. I will be happy to have a meeting with you and your **sales team** for ② the new products. I will be available next Tuesday and Wednesday evenings. Please let me know when is **suitable** for you. We can also schedule at ③ anytime the week after ④ if the above two evenings are not good for you. I will bring some test samples with me when I come for the meeting.

Looking forward to seeing you.

Sincerely,

Mike Rogers

嗨，珍妮佛：

很高興收到你的回信①。我很樂意和你及你的**行銷團隊**碰面談②新產品的事。我下週二和三的晚上都有空。請讓我知道你何時**方便**。如果上述兩個晚上的時間你都不行，我們也可以約③下下週④的任何時間。我去開會時會帶一些樣品過去的。

期待與你見面。

麥可．羅傑斯 謹上

換句話說

① **get your reply**

→郵件中的片語hear back from sb. 為固定用法，表示「收到某人的回覆」，而同義的表達方式還有get your reply，在此可以替換做使用。

② **to discuss / to talk about**

→郵件中的介系詞for後面可以直接帶出做某事的目的，也就是郵件中談論的meeting（會議），除此之外，也可以用替換語句to discuss或to talk about表示，更加強調「談論；討論」的動作。

③ **arrange for**

→郵件中的schedule為動詞，表示「將行程訂於……」，後接at和時間，除此之外，還可以用常用的arrange替換表示。值得注意的是arrange後接的介系詞為for，不可與schedule +at搞混。

④ **the week after next 下下週／after May 16 五月十六日以後／after the Dragon Boat Festival 端午節過後**

→郵件中的anytime the week after表示「下下週的任何時間」，除此之外，其他時間的表示方式也可以參考替換句：下下週the week after next、五月十六日以後after May 16、端午節過後after the Dragon Boat Festival等。

♥ **sales team** *n.* 行銷團隊
♥ **suitable** *adj.* 合適的
♥ **be glad to** 為……感到高興
♥ **hear back from sb.** 收到某人回信

馬上來練習吧！

想要索取試用品時該怎麼說呢？

1. 開頭稱謂 Dean,

2. 問候句 Thanks for ____________________.
I _______________, _______________,
_______________. _______________,
_______________.

3. 信件主要內容 Can you please ____________________
____
_________________________?

4. 結尾問候句 Thanks.

5. 署名 Howard Kieth

參考解答 Answer

Dean,

Thanks for the follow up email. I have been very impressed with your service so far, and also really like the samples. Right now I do not have questions about the product, I think you have it all covered already. I would, however, like to put in an order for 400 of these right away, as I think we could sell a lot. Can you please let me know how best to go about it?

Thanks.

Howard Kieth

中譯

迪恩：

感謝您的後續郵件。到目前為止，您們的服務讓我印象深刻，而且我非常喜歡這些樣品。目前我對產品並沒有什麼問題，我認為您們已經全部都處理得很好了。不過我想馬上訂400個，因為我想我們可以賣出很多。你能讓我知道你們能提供的最優惠價格嗎？

謝謝。

霍爾・齊耶思

句型這樣替換也可以

1. **Thanks for your new catalogue of...** 謝謝您……的新目錄
2. **I am glad to hear back from...** 很高興收到……的回信
3. **I will be available...** 我……會有空

Unit 03 詢問產品性能

寄信時可以這麼說

Dear George,

It has been a while ① since last time I was at your office. I hope everything is going well.

Last time you **mentioned** that you would have some of your customers **try out** the Brooks product samples ② that I brought you. Did you get any **feed back** from your customers? We would appreciate any feed back so that we can develop better products ③ to **meet** the customers' need ④.

Thanks.

Jane Rogers

親愛的喬治：

從我上次在辦公室至今已過了一段①時間了。希望您一切順利。

上次你**提到**您會請一些客戶**試用**我帶過去給您的布魯克斯樣品②，您有收到客戶給您的任何回應嗎？我們對任何**回饋**的意見都心存感激，如此我們才能開發更優良的產品③，以**迎合**顧客的需要④。

謝謝您。

珍・羅傑斯

換句話說

① **a long time**

→表示「有一陣子了」除了常會用it has been a while之外，也可以用替換句a long time表示。因此此句可以改寫為It has been a long time since last time I was at your office.。

② **our new line of cosmetics 我們新上市的化妝品系列／the trial version of our game design software 我們的遊戲設計軟體試用版**

→郵件中談及的是樣品product samples，其他種類的商品可以參考替換詞：我們新上市的化妝品系列our new line of cosmetics，和我們的遊戲設計軟體試用版the trial version of our game design software。

③ **improve our products 改進我們的產品**

→郵件中的develop better products表示「開發更優良的產品」，除此之外，也可以用「改進我們的產品」improve our products在此替換使用。

④ **to satisfy our customers 以滿足我們的顧客**

→郵件中的the meet one's need片語表示「迎合某人的需要」，換言之，就是想「滿足某人的要求」，因此此處可以用替換句th satisfy our customers表示。

♥ **mention** *v.* 談論；提及
♥ **feed back** *n.* 回饋
♥ **meet** *v.* 迎合
♥ **try out** 試用

回覆時可以這麼說

Dear Jane,

Very nice to hear from you.

We did do sample tests with some of our customers and **so far** ①, the users experience and feedback are all fairly ② **positive**. The only **concern** that some customers have is the prices ③ of some of the products. I will collect ④ all the information together and forward it to you. Hope this will help.

Sincerely,

Peter Frodo

親愛的珍：

很高興收到您的來信。

我們的確讓一些客人試用了樣品，**目前為止**①，使用者的經驗和回饋的意見都是非常②**正面**的。有些客人唯一**在意的**就是其中一些產品的價格③。我會將所有資料收集④起來，並轉寄給您。希望會有幫助。

彼得 · 佛羅多 謹上

換句話說

① **thus far / until now / to date**

→表示「到目前為止」除了可以用郵件中的so far以外，也可以用thus far、until now和to date替換表示。

② **quite / very / really**

→郵件中的fairly表示「相當地；非常」，除此之外，同樣意義還可以用其他副詞例如：quite、very或really表示。這類副詞都屬「程度副詞」，用以修飾形容詞的程度。

③ **the scents 氣味／the shape of the containers 產品容器的造型／the ingredients 成分**

→郵件中在此處提及的是產品的「價格」問題，除此之外，還可以參考替換詞句：氣味the scents、產品容器的造型the shape of the containers，和成分the ingredients，皆可在此作替換使用。

④ **gather 匯集**

→郵件中的collect表示「收集」之意，除此之外，還可以用gather一詞表示「匯集」。另外，gathering也是其常用的形式，作為名詞，表示「聚會；集會」之意。

♥ **positive** *adj.* 正面的
♥ **concern** *n.* 在意的事；關心的事
♥ **so far** 目前為止

馬上來練習吧！

想要詢問產品性能時該怎麼說呢？

1. 開頭稱謂　To Alice,

2. 問候句　Thanks for ____
____
____
__________.

3. 信件主要內容　I hope that ____________________________.
We really would like ____________________________
____
____________________.

4. 結尾問候句　Regard,

5. 署名　Warren Swan

參考解答 *Answer*

To Alice,

Thanks for volunteering to try out our new microwave. I hope that the installation was to your satisfaction. We really would like as much feedback as you can provide to us so that we can better address our future customers.

Regard,

Warren Swan

中譯

艾莉絲：

感謝您主動願意試用我們的新微波爐。希望這項設備能讓您滿意。我們真心希望您能盡可能的提供使用心得，好讓我們能夠使未來的客戶更為滿意。

華倫．史旺 謹上

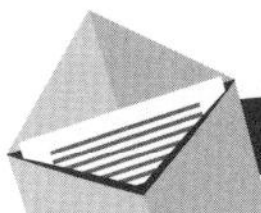

句型這樣替換也可以

1. **It has been a while since...** 自從……至今已過了一段時間了
2. **I hope everything goes well** 希望您一切順利
3. **Did you get any feed back from...** 您有收到……給您的任何回饋嗎

Unit 04 | 回覆產品問題

寄信時可以這麼說

Dear Jim,

We have **stocked the shelves with** the new product line and it has sold well ①. Speaking to ② my sales staff here, they have had a couple of comments from the consumers that you may be interested in. The most common comment has been that the product **works** much faster ③ than expected, and this has been good ④ feedback. The only **negative** comment so far has been about the **packaging** from a couple of older customers, who had **difficulty** with it.

Regards,

Mandy Stevens

親愛的吉姆：

我們已經**將**新產品系列**上架**，而且銷售得很好①。我和這裡的銷售人員談過②，他們有提到一些你可能會想知道的顧客意見。最普遍的意見就是產品的**效果**比原先預期的要來得快③，這應該是好的④回應。目前唯一的**負面**意見，是關於**包裝**；這是幾個對包裝有**異議**的老主顧提出來的。

曼蒂・史蒂文斯 謹上

換句話說

① **created an increasing sales volume 創造持續上升的銷量／become the best seller 成為熱銷商品**

→郵件中的(products) sold well表示「商品賣得很好」，除此之外，也可以用替代句「創造持續上升的銷量」created an increasing sales volume，或是「成為熱銷商品」become the best seller表示。

② **After I spoke to**

→郵件中此句以speaking開頭，並非是因為「談論」正在進行中，而是由after I spoke to...變化而來的「分詞構句」，用以簡化過於繁雜或冗長的句型。

③ **more effective 更有效率**

→faster一詞為fast得比較級，表示「更快的」，換言之，可以用「更有效率」表示，也就是替代詞more effective。值得注意的是，形容詞原形fast為單音節，所以在變化為比較級時，只需在其後方加上est；而形容詞effective則因有較多音節，所以要另外在其前面加上more一詞，才能變為比較級的形式。

④ **possitive 正面的**

→在英文中，通常表示「正面的」會用possitive一詞，其為郵件後面使用的negative（負面的）的反義詞，通常會一起搭配使用，因此此處可以將good改為possitive。

單字片語急救包

- ♥ **work** *v.* 起作用
- ♥ **negative** *adj.* 負面的
- ♥ **packaging** *n.* 包裝
- ♥ **difficulty** *n.* 異議；反對困難；難題
- ♥ **stocked the shelves with** 將……貨品上架

回覆時可以這麼說

Dear Mandy,

Thank you so much for your feedback. It is comments like these that have helped us become a leader ①in the industry. ② I will be **sitting down with** the **product development group** next week and will **pass** all of your feedback **along**. I'll certainly ③ **bring** this **up** in the meeting and let you know what we plan to do ④.

Thanks again.

Jim Hart

親愛的曼蒂：

非常感謝您的意見。就是有像這樣的意見，才能幫助我們成為業界的領導者①。②下週，我會和**產品開發小組坐下來開會**，並將所有您回饋的意見**讓每個人知道**。我一定③會在會議中將這件事**提出**來，並讓你知道我們計畫怎麼處理④。

再次感謝你。

吉姆・哈特

換句話說

① a leading company / a giant

→表示某公司為該產業中的「龍頭」時，除了可以用郵件中的leader以外，也可以轉換其詞性，改用a leading company表示，或是直接使用giant也可以。

② The comments like these always help us to maintain our first place in the industry.

→郵件中的It is是虛主詞，用以代替後面提到的the comments the these，有強調的功用。除此之外，也可以用一般句型The comments like these開頭，因此此句可以改寫為The comments like these always help us to maintain our first place in the industry.。

③ definitely / make sure to / ensure to

→郵件中的certainly表示「一定會（做某事）」，也可以用同義的副詞definitely替代表示。除此之外，也可以使用make sure to或ensure to兩個片語表示「確認會（做某事）」。

④ our next step 我們的下一步

→在英文中，有時會用next step表示之後的打算、動作，所以此處可以將原句替換為...let you know our next step（讓您知道我們的下一步）。

單字片語急救包

♥ **product development group** *n.* 產品開發組

♥ **sit down with...** 與……一起坐下來，引申為「坐下來開會、討論」的意思
♥ **pass...along** 將……一個接一個得傳遞
♥ **bring...up** 提出

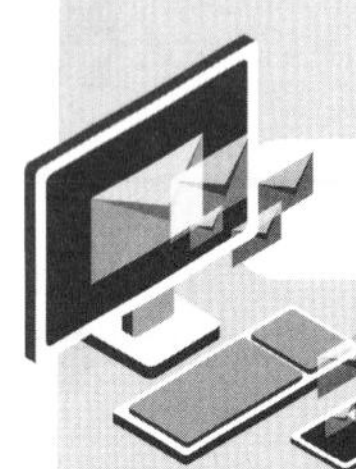

馬上來練習吧！

想要回覆產品問題時該怎麼說呢？

1. 開頭稱謂　Wareen,

2. 問候句　Thanks again for ____________________ ____________________.

3. 信件主要內容

The installation was ____________________ ____________________.

I must admit that ____________________ ____________________.

Could you maybe ____________________ ____________________?

4. 結尾問候句　Thanks.

5. 署名　Alice Barren

參考解答 Answer

Warren,

Thanks again for letting me be part of your development of this great product. The installation was top notch and the installer gave me some really good tips on how to use it. I must admit that the manual is very large and a little scary. Could you maybe include a shorter cheat sheet so that first time users can get cooking straight away?

Thanks.

Alice Barren

中譯

華倫：

再次感謝您讓我參與這項了不起的產品的開發。這項設備是頂級的，而且安裝人員給我一些很好的使用技巧。我必須說，使用手冊實在很大本，而且有點嚇人。你們或許可以附上一張短一點的小抄，如此一來第一次使用的人就可以直接烹調了。

謝謝您。

艾莉絲・拜倫

句型這樣替換也可以

1. **We would appreciate any feed back so that...** 我們對任何回饋的意見都心存感激，這樣我們才能……
2. **Hope this will help** 希望會有幫助
3. **Thanks for the prompt shipment of...** 感謝您將……快速送達

Unit 05 寄出貨物通知

寄信時可以這麼說

Hi Jeff,

We have processed your **purchase order** # 25513 and the order is shipped out today by **Fedex Ground**.

You should expect the package to arrive ① at your office within 3-4 business days ②.

If you have any further questions, please feel free to ③ contact us.

Again, thanks for your support of our business ④.

Sincerely,

Tamara Hopkins

嗨，傑夫：

我們已經處理了你們編號25513號的**訂購單**， 你們的貨品今天已經以**聯邦路面快遞送過去了**。

包裹應該在三到四個工作天內②就會抵達①你們公司 了。

如果有任何問題，請不吝③聯絡。

再次謝謝您們對我們公司的支持④。

塔瑪拉 · 霍普金斯 謹上

換句話說

① the arrival of the package

→動詞arrive表示「到達」，arrival則為其名詞形式，因此此處可以以名詞形式改寫為the arrival of the parckage。

② within the next two days 未來兩天之內／within the week 在本週之內

→within一詞表示「在……（某時間）內」，因此郵件中的within 3-4 business days表示「三到四個工作天」，另外，還可以用替換句「未來兩天之內」within the next two days，或是「在本週之內」within the week等表示。

③ don't hesitate to 不要猶豫……

→在商業書信的溝通中，常會告知對方「有問題可以盡量提出」，而郵件中的feel free to contact us便是此意。除此之外，常見的說法還有don't hesitate to contact us，表示「不用猶豫，直接向我們聯絡」之意。

④ thank you for supporting our business

→郵件中的thank for one's sth.可以在句型上稍作改變，成為thank sb. for Ving，因此此句可以改寫為thank you for supporting our business。

♥ **purchase order** *n.* 訂購單
♥ **Fedex Ground** *n.* 聯邦快遞路面運輸

回覆時可以這麼說

Dear Tamara,

Thanks for the update. I'll expect it in 3-4 days. Would it be possible to **send through** the **waybill** number for our records ① **in case** ② the shipment does not arrive on the expected day? That way ③ I can **trace** it myself and won't need to **trouble** you again ④.

Many thanks for your excellent service.

Yours sincerely,

Jeff Nixon

親愛的塔瑪拉：

謝謝您告訴我最新的訊息。這三到四天之間，我會等貨到來。是不是可以請您將**運貨單**號碼**告訴**我以供紀錄①，**以免**②貨物沒有在預期的那天送來。如此一來，我就可以自行**追蹤**貨物，而不需要再次④**麻煩**你。

感謝您優質的服務。

傑夫・尼克森 謹上

換句話說

① for us to keep as a record

→for one's record表示「以供……紀錄」，完整寫出則為...for us to keep as a record，因此可以在此作替換使用。

② lest

→郵件中的in case表示「以免……」，同義字詞可以使用lest作為替換使用。值得注意的是，lest使用的句型為lest + S + (should) + V，因為動詞前原有should，所以就算省略掉了，動詞也要使用原形動詞。

③ Thus / Therefore

→郵件中的that way是代替前面一句提到的send though the waybill，表示「如此一來」。另外，此處也可以直接用thus和therefore替換表示。

④ once more

→郵件中的again表示「再次」，還可以用onc more替換表示。其中的once表示「一次」，因此加上more便也作「再一次」。

單字片語急救包

♥ **waybill** *n.* 運貨單
♥ **trace** *v.* 追蹤
♥ **trouble** *v.* 麻煩
♥ **send through** 通知；報告
♥ **in case** 以免

馬上來練習吧！

想要表示寄出貨物通知時該怎麼說呢？

1. 開頭稱謂　To Caltech Engineering,

2. 問候句　Thank you ______________________.

3. 信件主要內容

Your order for ______________________

______________________.

____

______________________.

Please ______________________

______________________.

4. 結尾問候句　Regards,

5. 署名　William Young

參考解答 *Answer*

To Caltech Engineering,

Thank you for your order. Your order for 5 boxes of cable has been shipped today from our Seattle warehouse. The shipping number is 43532832 and it has been shipped by FedEx. Please feel free to follow this online.

Please keep us informed regarding you need additional orders.

Regards,

William Young

中譯

你所訂購的5箱電纜已於今天從我們西雅圖的倉庫送出。貨運單號為43532832，交由聯邦快遞運送。請您直接上網追縱貨物。

若您需要追加訂單，請與我們聯絡。

威廉・楊 謹上

句型這樣替換也可以

1. **The order is shipped out today by...**
   貨品已於今天以……運送過去了
2. **You should expect the package to arrive at...within...**
   包裹應該在……內就會抵達……了
3. **Would it be possible for me to...** 我是不是可以……
4. **Many thanks for...** 感謝您……
5. **If there are any other concerns, please...**
   若有任何其他重要事項請……

Unit 01 | 請求廠商提供零件資料

寄信時可以這麼說

Dear Arnold,

As we are running ① some parts of a new project, we will be needing **a series of** ② graphics cards ③ to fit our existing **server** array which we purchased from your company last summer. Would you please let me know ④ the **availability** and **cost** of these cards?

Your prompt reply is highly appreciated.

Best Regards,

Martha Vader

親愛的亞諾：

我們正在進行①一個新案子的一部分，會需要**一系列**②能夠適用於我們現有的，亦即我們去年向你們公司購買的**伺服器**陣列的繪圖卡③。能不能麻煩您告訴我④這些繪圖卡的**可得性**以及**費用**呢？

若能快速回覆我，將感激不盡。

瑪莎・威德 謹上

換句話說

① we have begun to proceed with

→郵件中的running表示「進行中的」，也可以用proceed with片語替代表示。若請人繼續發言，也可以直接說 Please proceed.（請繼續。）。

② a set of 一組

→郵件中的a series of表示「一系列」，此處也可以用a set of表示「一組」，雖然在意義上兩者有些微不同，但是在此處使用是大同小異的。

③ video cards 顯示卡／central processors 中央處理器／operating system 作業系統

→郵件中提及的產品為繪圖卡graphics cards，其他相關的產品可以參考替換詞：顯示卡video cards、中央處理器central processors和作業系統operating system等。

④ May I ask you about... 我能否向您詢問……

→郵件中的最後一句would you please...是以對方的角度切入的詢問句，表示「您是否能……」，而若使用替換句May I ask you about...則式表示「我能否詢問您……」。值得注意的是，may只會用於詢問對方自己是否能做某事之時，所以不能說May you do sth.。

♥ **server** *n.* 伺服器
♥ **availability** *n.* 可得性
♥ **cost** *n.* 費用、成本
♥ **a series of** 一系列

回覆時可以這麼說

Martha,

Great to hear from you again. I hope the servers are still all running ① **flawlessly** ②. For the servers you have, we can either use 512MB or 1GB cards. These are $170 and $400 **respectively**. We presently ③ only have seven 512MB cards **in stock** and no 1GB ones; however, I can get them **at short notice** ④.

Looking forward to helping you.

Arnold Lakshemeer

瑪莎：

很高興接到您的來信。希望所有的服務器都仍然**毫無瑕疵地**②運作①中。針對您們的伺服器，我們可以用容量512MB或是1GB的顯示卡。這些**分別**是170美金和400美金。目前③我們**現貨**只有七個512MB的顯示卡，沒有1GB的，不過只要您**一通知**我，我**就可以**④拿貨。

期待能夠協助您。

亞諾．拉克許米爾

換句話說

① operating

→除了郵件中的動詞run可以用來表示運作之外，operate也同樣可以在此作替換使用。另外，operate的名詞形式operation除了表示公司的營運狀況外，也常用於醫療手術之時。

② without any problem

→郵件中的flawlessly表示「完美地」，為形容詞flawless「沒有缺點的」變化而來，此外，在此還可以用替換句without any problem表示「沒有任何問題」。

③ currently / now

→presently一詞表示「目前；當下」，因此在此也可以用副詞currently替換表示，或是用now直接表示「現在」之意。

④ once I am informed

→郵件中的in a short notice表示「一接到通知就……」，而同樣可以表示「一……就……」的句型還有once可以使用，因此此處可以改寫為I can get them once I am informed。

♥ **flawlessly** *adv.* 完美無瑕地
♥ **respectively** *adv.* 分別地；各自地
♥ **in stock** 有庫存
♥ **at short notice** 一接到通知就……

馬上來練習吧！

想要請求廠商提供零件資料

1. 開頭稱謂　George,

2. 問候句　We ________________. ________________.

3. 信件主要內容

Do you have any of the following items -________
____
____
____

4. 結尾問候句

Please ________________________________
____
____________.

Thanks.

5. 署名　Peter Salmon

參考解答 Answer

George,

We are embarking the new motor project that we spoke about last week. At that time, I mentioned that we may need your help to source some parts.

Do you have any of the following items:

A 7Amp winder coil

A couple of 3.4mm bearings

5.6 or 5.8mm low profile fans

Please let me know ASAP so I we can get them in if you do.

Thanks.

Peter Salmon

中譯

喬治：

我們目前正全力進行上週我們談到的發動機新案子。當時我有提到，我們可能需要您協助採購某些零件。

您有沒有以下任何一種物品：

一個7Amp的捲線機

一些三點四毫米的軸承

5.6或5.8厘米的低調風扇

如果你們有的話，請立刻讓我知道，以讓我進貨。

謝謝。

彼得・沙蒙

句型這樣替換也可以

1. **We are in great need of...** 我們急需……
2. **Would you please let me know...** 能不能麻煩您告訴我……

Unit 02 詢問專案研發內容

寄信時可以這麼說

Dear Kamal,

It has been a couple months since we last spoke and I was wondering ① how the new traffic light ② project is going. I am pleased to let you know that we have additional ③ **poles** ④ in stock, at a lower price than the original ones we shipped to you.

Please **drop** me **a line** and I can **fill** you **in** on the details.

Thanks.

Jill Silousous

親愛的凱莫：

自我們上回談話至今已有好幾個月了。我想知道①那個新的交通信號燈②的案子進行得如何。很樂意通知您，我們現在已經有更多的③**柱子**④了，而且價格比原來我們運給你們的那一批還要便宜。

請**寫封信給**我，我會**告訴**你詳細內容。

謝謝。

吉兒・希勒瑟斯

換句話說

① **am curious about 對……感到好奇**

→郵件中的be wonder表示「想知道；有疑問」，此處還可以用be curious about表示「對……感到好奇」，雖然在意義上有所差異，但此處使用上並沒有太大差別。

② **house redecorating 房屋重新裝潢／overhead projector 投影機／street light 路燈**

→郵件中談論的案子是與交通信號燈traffic light有關，其他相關案子的寫法可以參考替換詞：房屋重新裝潢house redecorating、投影機overhead projector，和路燈street light等。

③ **extra 另外的；外加的**

→郵件中的additional表示「更多的」，除此之外，還可以用extra一詞表示「另外的；外加的」。值得注意的是，通常extra會用瑜表示「額外」有的東西，例如：Do you have an extra pen which I can borrow?（你有沒有多的筆可以借我？）。

④ **porcelain tiles 瓷磚／projector screens 投影機螢幕／LED light bulbs 二極光燈泡**

→郵件中提論的產品為柱子pole，其他產品的寫法可以參考替換詞：瓷磚porcelain tiles、投影機螢幕projector screens，和二極光燈泡LED light bulbs等。

♥ **pole** *n.* 柱子

♥ **drop sb. a line** 寫信給某人

♥ **fill sb. in** 通知

回覆時可以這麼說

Jill,

I appreciate the follow-up. Unfortunately ① we have **hit** some problems on that project with low temperature performance ② and will need to do a **redesign**. This may also mean ③ we might need to **modify** the pole **assembly**. I'll let you know once ④ the engineers have finished the additional testing.

Regards.

Kamal

吉兒：

很謝謝你的後續行動。遺憾的是①，我們在這個案子上的低溫性能方面②**遇到**了一點問題，將必須**重新設計**。而這也許意味著③我們必須**修改**燈柱的**裝配**。只要工程師完成進一步的測試，我就會④馬上通知你。

祝 好。

凱莫

換句話說

① Sadly

→郵件中的unfortunately表示「不幸的是；遺憾的是」，除此之外，還可以用副詞sadly替換使用於此。

② red light camera devices 闖紅燈自動照相裝置／speed camera devices 超速自動照相裝置

→郵件中提及的問題為「低溫性能」low temperature performance，其他與交通信號燈相關的問題還有：「闖紅燈自動照相裝置」red light camera devices，和「超速自動照相裝置」speed camera devices等，可以在此作替換使用。

③ suggest

→suggest一詞除了表示「提議；建議」以外，也表示「代表；意味」，所以郵件中的mean可以替換為suggest使用。

④ as soon as

→once一詞表示「一……就……」，同樣字義的表達方法還有as soon as...，因此此句可以改為I'll let you know as soon as the engineers have finished the additional testing.。

- ♥ **hit** *v.* 碰；遭到
- ♥ **redesign** *v.* 重新設計
- ♥ **modify** *v.* 修改；改造
- ♥ **assembly** *n.* 裝配；組裝

馬上來練習吧！

回覆詢問專案研發內容時該怎麼說呢？

1. 開頭稱謂　Peter,

2. 問候句　Thanks ________________________________
____
______________________________________.

3. 信件主要內容　I ______________________________________.
______________________________________.
____
____
____________.

4. 結尾問候句　Looking forward to your reply.

5. 署名　George Stevens

參考解答 Answer

Peter,

Thanks for the email. I can certainly help you out with the bearings and the fans, although we only carry 5.8mm ones. As far as the winder coil goes, I am going to need some more information from you. We will need to look at the performance parameters and dimensions you require to choose one.

Looking forward to your reply.

George Stevens

中譯

彼得：

謝謝您的來信。我可以在軸承和風扇這方面提供你協助，儘管我們只有5.8厘米的。至於捲線機，我可能需要你提供更多的資訊。我們需要依照你們所需要的性能參數和規模來選擇。

期待您的回覆。

喬治・史蒂文斯

句型這樣替換也可以

1. **Your prompt reply is highly appreciated** 快速回覆，感激涕零
2. **We have not heard any feedback on...** 我們一直沒有聽到任何有關……的回饋

Unit 03 | 要求產品性能解釋

寄信時可以這麼說

George,

I see that you have recently released the third generation mouse ① for general production. The sales team and I have spent some time ② with this **device** to see how best ③ to promote it to our customer base. Unfortunately, we have **run into** a lot of questions on the product, and have even tried ④ to use it ourselves **with little success**. Could you **take** the team **through** the features and abilities of this mouse when time allows?

Sincerely,

Anna Johnson

喬治：

我知道貴公司最近已經發佈要普遍生產第三代滑鼠①。我和行銷團隊花了一些時間②使用這個**裝置**，想研究出該如何最有效地③將之推銷給我們的顧客群。很遺憾的是，我們在這個產品上**遇到**很多問題，甚至我們已經試著④自己使用，但**成效甚微**。如果時間允許的話，您能不能**向**團隊**說明**這個滑鼠的特點和功能呢？

安娜·強森 謹上

換句話說

① **smart phone 智慧型手機／laptop 輕便型電腦／cordless keyboard 無線鍵盤**

→郵件中提及的產品為滑鼠mouse，其他相關產品的寫法可以參考替換詞：智慧型手機smart phone、輕便型電腦laptop，和無線鍵盤cordless keyboard等。

② **It takes me and the sales team some time**

→同樣表示「花費（時間；金錢）」的方式主要有二。一為「人 + spend + 時間」，也就是郵件中原句的寫法；二則為「It + takes + 人 + 時間」，或是「做某事 + takes + 人 + 時間」，也就是替換句的It takes me and the sales team some time...。值得注意的是，以spend做動詞時，主詞只能為人；而以take做動詞時，主詞則不能為人。

③ **to find the best way**

→郵件中的how best兩字直接表達「如何」和「最好地」，因此整句意義為「如何最有效地……」。除此之外，還可以用to find the best way to...替換此句。

④ **attempt**

→表示「嘗試；試著」除了可以用常見的動詞try以外，還可以使用替換動詞attempt表示。另外，attempt也可以做名詞使用。

♥ **device** *n.* 裝置
♥ **run into** 遇到
♥ **with little success** 難有所成
♥ **take sb. through** 對某人講解；帶領某人做……

回覆時可以這麼說

Dear Anna,

We did **extensive** testing on the product and really feel ① that it will fill a definite **niche** that no other products ② presently ③ cover. This is best explained **by means of** a **demonstration**, and I would love to provide one to the team. Also please note ④ that most of the **functionality** for the mouse comes after **installation** of the **software drivers**.

Thanks.

George

親愛的安娜：

我們針對這個產品做了**大規模**的測試，而且真的認為①它能達到一定的**利基**，是目前③其他產品②無法做到的。這要**用**實地示範**的方法**才能做最佳解釋。我很樂意向團隊提供一次**示範**。還有，請注意④這個滑鼠的功能要在**安裝軟體驅動程式**之後才能發揮到最大。

謝謝。

喬治 謹上

換句話說

① agree

→動詞agree可以表示「一致認為；一致同意」，因此在此可以做替換使用，正句改為We did...and agree that it will...。值得注意的是，agree的用法除了在其後加上子句，也可以用片語agree on sth.。

② nothing else

→no other的用法表示「沒有其他……」，用法很廣泛，例如：No other student in the class is smarter than Vivian.（班上沒有其他人比薇薇安聰明）。除此之外，還可以用nothing else表示。值得注意的是，else後方不用加上名詞products，相關用法還有：who else、what else、anything else、where else等。

③ currently / now

→郵件中的presently表示「目前；現在」，同義的副詞還有currently和now可以在此做替換使用。

④ be reminded

→please note that...在英文中是常用的句型，用以提醒某人某件事，除此之外 ，還可以用please be reminded片語，同樣表示「請注意；請留意」之意。

單字片語急救包

- ♥ **extensive** *adj.* 廣泛的
- ♥ **niche** *n.* 利基
- ♥ **demonstration** *n.* 實地示範
- ♥ **functionality** *n.* 功能
- ♥ **installation** *n.* 安裝
- ♥ **software driver** *n.* 軟體驅動程式
- ♥ **by means of** 用……；以……

馬上來練習吧！

回覆要求產品性能解釋時該怎麼說呢？

1.	開頭稱謂	Ron,
2.	問候句	Thanks ______________________________ ______________________________ ______.
3.	信件主要內容	In terms of ______________________________, ______________________________. I do feel that ______________________________, ______________________________ ______________________________.
4.	結尾問候句	Thanks.
5.	署名	Jim Ankle

參考解答 Answer

Ron,

Thanks for the pens. They have been great. In terms of actual writing, I think they are fine; however I do have a comment. Given that we are calling these “platinum” and the price point we are aiming at, I do feel that the pen itself feels very light in the hand. The build quality is fine, but I would have thought that we could make the pen feel a little heavier and more substantial.

Thanks.

Jim Ankle

中譯

榮恩：

感謝您提供的筆，他們真的是很好用。在實際書寫方面，我認為他們是很好，但我的確有些意見。鑑於我們稱之為「白金」，而且價格也道出我們的目標，我實在覺得這筆握在手中感覺很輕。構造品質是好的，但我認為，我們可以讓筆身更有重量、更覺得有點重，更實在一些。

謝謝您。

吉姆・安可

句型這樣替換也可以

1. **I would like to have a round table discussion on...**
   我想在……開個圓桌會議
2. **There are quite a number of...** 我們有很多……
3. **I will send...to you later on** 我稍後將會寄……給您
4. **Would...suit you** 請問……您方便嗎
5. **I see that...** 我知道……
6. **I was interested to hear about...** 聽到……讓我很感興趣

Unit 04 | 請求技術協助

寄信時可以這麼說

To Tech Support,

We recently ① installed 5 of your printers in our network and are seeing **consistent** issues with them. **In summary** ②:

- All printers have formatting issues ③ when used from Microsoft Outlook.

- We see regular ④ **disconnects** of the printers from the print server.

Your prompt response is greatly appreciated.

Andrew Lopez

致技術支援部：

我們最近①安裝了五台你們的印表機在我們的網路上，發現它們有**一致的**問題。**摘要如下**②：

- 所有的印表機在使用 Microsoft Outlook程式時會出現格式化問題③。

- 我們發現印表機和印表機伺服器會經常性的④**斷線**。

您若能快速回覆，將感激不盡。

安德魯．羅佩茲

換句話說

① **lately**

→表示「最近」除了可以使用郵件中的recently以外，也可以用替代詞lately副詞。

② **in brief**

→summary一詞表示「總結；摘要」，因此郵件中的in summary表示「結論而言；概括而言」，同義片語in brief也可用於此。通常而言，這兩個片語常常會出現於學術寫作中的最後一段，帶出整篇文章的結尾。

③ **quit unexpectedly 閃退／stop working 停止運作**

→郵件中提及的機器問題為「格式化問題」formatting issues，另外相關的機器問題寫法可以參考替代詞：閃退quit unexpectedly和停止運作stop working，也可以在此作替換使用。

④ **constant / frequent**

→郵件中的regular表示「經常性的」，除此之外，也可以使用替換詞constant或frequent表示。

♥ **consistent** *adj.* 一致的
♥ **disconnect** *n.* 斷線；分離
♥ **in summary** 整體而言

回覆時可以這麼說

Dear Andrew,

Thanks you for your request ① for support. Can you please confirm the **model number** of the printers in question as well as ② the drivers you are using? The issues you are experiencing ③ all sound like they ④ relate to a **bug** in the 1.0.34 driver. Please upgrade this by **downloading** the latest driver from here: http://drivers.printers.net

Thanks.

Printer Support Team

親愛的安德魯：

謝謝您來函請求支援①。能不能請你確認有問題的印表機**型號**和②你們使用的驅動程式呢？你們遇到的③問題聽起來④跟1.0.34這個驅動程式中的**錯誤程式**有關。請到這個網址**下載**最新的驅動程式：http://drivers.printers.net

謝謝。

印表機支援組

換句話說

① requesting / asking

→Thank sb. for片語的使用，除了可以在其後加上one's N.以外，例如：Thank you for your help，也可以用thank sb. for Ving的形式，也就是替換句的用法Thank you for requesting / asking for support。

② and

→片語as well as表示「和」，用以連接前後兩個對等的詞句，等同於連接詞and，因此可以在此做替換使用。

③ you mentioned

→郵件中的you are experiencing為代替原寄信者所提及的兩的問題點，在此用以避免重複冗長的敘述，同樣的作用，還可以使用you mentioned可以一言蔽之，因此此句可以改寫為the issues you mentioned all...。

④ seem to be

→在英文中，通常表示「好像；似乎」等語氣時，和中文一樣習慣用「聽起來」和「看起來」，分別也就是郵件中的sould like和替代句的seem to be，因此在使用上通常可以互相代替。

單字片語急救包

♥ **model number** *n.* 型號
♥ **bug** *n.* 錯誤程式
♥ **download** *v.* 下載

馬上來練習吧！

想要請求技術協助時該怎麼說呢？

1. 開頭稱謂　To the Support Team,

2. 問候句　I am writing to ______________________.

3. 信件主要內容

We have just started using ________________.

______________________________________,

______________________________________.

______________________________________,

________________.

4. 結尾問候句　Looking forward to your help.

5. 署名　Christopher Robinson

參考解答 Answer

To the Support Team,

I am writing to talk about some problems we've encounter. We have just started using one of your 500 horse power motors at a factory in Chicago, and have run into overheating issues. Basically the motor starts and runs fine for about an hour and then cuts out. The display shows a temperature warning. If we leave it for half an hour it will fire up again, but will run into the same issue.

Looking forward to your help.

Christopher Robinson

中譯

致支援小組：

我們剛開始在芝加哥廠使用你們500馬力的電動機，並遇到過熱的問題。基本上電機啟動後會正常運作約一小時，然後就自動關掉。螢幕會顯示溫度警告。如果我們把它放著半小時，它就會再度啟動，但是同樣問題又會再發生。

期待您的協助。

克里斯多夫・羅賓森

句型這樣替換也可以

1. **Thanks you for your request for...** 謝謝您來函請求……
2. **Can you please confirm...** 能不能請你確認……
3. **If you have any further question, please quote...** 如果您還有任何疑問，請引用……

寄信時可以這麼說

Dear Charles,

Following ①the purchase by your company of our **nexus accounting software** ②, we have not heard any feedback ③ on the installation and **commissioning process**. Our team is here to help, so please feel free to ④ ask any questions about getting this software up in your environment.

Thanks.

Alison Swenson

親愛的查爾斯：

自①您向我們購買**一系列**會計軟體②之後，我們一直沒有聽到任何有關在安裝和**執行程序**上的反應③。我們的團隊在此提供協助，請不吝④諮詢任何有關使用此軟體的問題。

感謝您。

艾莉森．史文森

換句話說

① Since

→郵件的首句以following作為開頭，表示「自從……後」，同樣可以用since作為替換使用。值得注意的是，此類表示「從某時間一直延續表現在的」詞，接續的下半句必須使用現在完成式，也就是郵件中的have not heard...。

② graphic software 繪圖軟體／word processing software 文書處理軟體

→郵件中提及的產品為「會計軟體」accounting software，其他相關的產品包刮「繪圖軟體」graphic software和「文書處理軟體」word processing software可以在此作替換使用。

③ observation / evaluation

→郵件中的feedback表示「回饋；反應」，除此之外，還可以用observation和evaluation表示。

④ don't hesitate to

→郵件中的feel free to...用以邀約對方作某事，同樣的意義，也可以用don't hesitate to...表示，因此此處也可以改寫為please don't hesitate to ask any questions about...。

單字片語急救包

♥ **nexus** *n.* 一系列；一組
♥ **commissioning process** *n.* 執行程序

回覆時可以這麼說

Dear Alison,

Thanks for the follow-up. Unfortunately ①, due to ② **the length of time** it took to **conclude** the transaction **with** your organization, we have had to go to an alternative **open source package** in the meantime ③. This package **meets** all our requirements ④ at present, so we do not need the ongoing support contract you mentioned in a previous email.

Thank you very much.

Sincerely,

Charles Campbell

親愛的艾莉森：

感謝您的後續服務。可惜的是①，由於②我們花在與貴組織**完成**交易的**時間**太長，我們同時間③不得不去找一個可替代的**開放原始碼套裝軟體**。這套軟體**符合**我們目前所有的需求④，所以我們不需要您在上一封電子郵件中提及的持續支援合約。

非常感謝您。

查爾斯・坎貝爾 謹上

換句話說

① Sadly / However

→郵件中的unfortunately表示「不幸的是」，同義的副詞還有sadly可以在此做替換使用。除此之外，也可以用轉接詞however直接表示語氣上和談論內容的轉換。

② owing to / because of

→郵件中的due to表示「因為；由於」，同義的片語還有owing to和because of可以替換使用。值得注意的是，這三個片語後只能接名詞，若要接子句則要使用because表示。

③ at the same time

→郵件中的in the meantime表示「同時」，同義的片語還可以使用at the same time表示。不過相較之下，前者會比後者更加適合用於書信寫作。

④ needs

→表示「需求」除了使用郵件中的requirement以外，也可以直接用need表示，通常表示「符合某人期望」也會用meet one's needs或meet one's expectation表示。

單字片語急救包

- ♥ **open source** *n.* 開放原始碼軟體
- ♥ **package** *n.* 套裝軟體
- ♥ **meet** *v.* 符合
- ♥ **the length of time** 時間長度
- ♥ **conclude sth. with sb.** 完成與某人之間的某事

馬上來練習吧！

想要提供後續服務時該怎麼說呢？

1. 開頭稱謂　Dear Christopher,

2. 問候句　Thanks for ____________________.

3. 信件主要內容

I have seen ____________________.

______________________________.

Please ______________________________.

If you still have issues ____________________

____________________.

4. 結尾問候句　Thanks.

5. 署名　The Motor Support Team

參考解答 Answer

Dear Christopher,

Thanks for your inquiry. I have seen this issue a number of times and it is normally caused by not removing the transportation protector upon installation. If you look at the back of the motor, there is a yellow plate that says please remove before use. Please remove this plate and keep it in a safe place in case you need to move the motor again. If you still have issues after this, please do not hesitate to contact us.

Thanks.

The Motor Support Team

中譯

克里斯多夫：

感謝您的諮詢。我看過這個問題很多次，而它通常是因為沒有在安裝時取消運輸保護裝置所導致。如果你看一下電動機的背後，會看到一個黃色牌子，上面寫著使用前請移除。請您移除這塊牌子並將之置於一個安全的地方，以防您將來需要再次移動電動機。如果您仍然有問題，請儘管與我們聯繫。

謝謝。

電機支援組

句型這樣替換也可以

1. **Please feel free to ask...** 請不吝諮詢……
2. **I was wondering if you could...** 我想知道您是否能……
3. **I would appreciate if you could...** 如果您能……我將非常感激

原來如此 系列 E219

深度解密！一次就上手的超實用職場英文 E-mail 即戰手冊

豐富的E-mail 情境式主題，讓你關關突破職場上的溝通問題！

作　　者　鍾君豪◎著
顧　　問　曾文旭
社　　長　王毓芳
編輯統籌　耿文國、黃璽宇
主　　編　吳靜宜、姜怡安
執行主編　李念茨
執行編輯　陳儀蓁
美術編輯　王桂芳、張嘉容
封面設計　阿作
法律顧問　北辰著作權事務所　蕭雄淋律師、幸秋妙律師

初　　版　2020年01月
出　　版　捷徑文化出版事業有限公司
電　　話　（02）2752-5618
傳　　真　（02）2752-5619
地　　址　106 台北市大安區忠孝東路四段250號11樓-1

定　　價　新台幣350元／港幣 117 元
產品內容　1書

總 經 銷　采舍國際有限公司
地　　址　235 新北市中和區中山路二段366巷10號3樓
電　　話　（02）8245-8786
傳　　真　（02）8245-8718

港澳地區總經銷　和平圖書有限公司
地　　址　香港柴灣嘉業街12號百樂門大廈17樓
電　　話　（852）2804-6687
傳　　真　（852）2804-6409

▶本書部分圖片由 Shutterstock、freepik 圖庫提供。

捷徑Book站

國家圖書館出版品預行編目資料

深度解密！一次就上手的超實用職場英文 E-mail 即戰手冊／鍾君豪著. -- 初版. -- 臺北市：捷徑文化, 2020.01
　面；　公分
ISBN 978-957-8904-95-8(平裝)

1.商業書信 2.商業英文 3.商業應用文 4.電子郵件

493.6　　108014385